Thermal Insulation
and Condensation

Thermal Insulation and Condensation

Paul Marsh

THE CONSTRUCTION PRESS

LANCASTER   LONDON   NEW YORK

The Construction Press Ltd,
Lancaster, England.

A subsidiary company of Longman Group Ltd, London.
Associated companies, branches and representatives
throughout the world.

Published in the United States of America by
Longman Inc, New York.

First published 1979

ISBN 0 860958 14 0

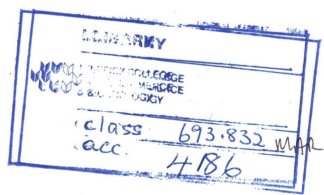

Printed in Great Britain at The Pitman Press Bath

Contents

Acknowledgements

The author acknowledges with thanks the help and advice given by the following during the preparation of the book:

N O Milbank	Building Research Establishment
J Peach	Chartered Institution of Building Services
D P Turner	Pilkington Brothers Limited

In addition he gratefully acknowledges permission to include in this volume tables and figures derived from other publications and sources. The organsiations and their material used are as follows:

British Standards Institution.
Figures 4.8, 5.10 and 6.3.

Chartered Institution of Building Services.
Figures 2.2, 4.4, 4.9 and 4.10,
Tables 4.1, 4.2, 4.3, 4.5, 4.6 and 4.7.

Pilkington Brothers Limited.
Figures 4.12 and 7.5.

BRE/HMSO
Figures 2.1, 3.1, 4.1, 4.3, 5.11 and 6.1.
Tables 1.1, 3.1, 3.2, 3.3, 4.8, 5.3, 5.5, 5.6, 5.8, 5.9, 5.11, 5.12, 5.13 and 5.14 appear by courtesy of the Director, Building Research Establishment. They are reproduced by permission of the Controller, H.M.S.O. and are largely derived from the Crown copyright BRE Current Papers CP 5/72, CP 2/74, CP 75/76 and CP 64/76, and from BRE Digests 108, 110, 119, 140, 145 and 210

Introduction

As the appreciation has grown that fossil fuels are a limited resource, an appreciation emphasised by exhortations from the government to 'save it' and by penal price rises affecting every aspect of our economy, so the thermal insulation of our buildings has become a vital factor in the way we construct them.

Energy is now recognised as a valuable commodity, not to be frittered away through draughty windows, up unused flues, or even through poorly insulated walls and roofs. We are encouraged to seal up every draughty crack, to double-glaze windows, foam-fill wall cavities, insulate roof spaces. Insulation has become the big 1970's trend, backed by all the devices of the present-day marketing machine. But how much of this extra expense is really reflected in substantial energy saving? Where can our money be best spent to show a real reduction in our electricity or gas bill? Are we, maybe, merely creating more congenial living conditions, more compatible with new patterns of life, while spending the same amount on fuel? How many years will it be before the capital cost of separate pane double-glazing is recouped in fuel savings – and will this happen before the timber windows, to which the aluminium secondary frames are fixed, have fallen to pieces?

This book (which is the second in a series by the same author dealing with the performance of the building shell[1]) endeavours to take an objective view of the current situation, trimmed of its marketing propaganda. With the help of much as yet unpublished research it hopes to assess the best methods of designing our buildings so that they are thermally efficient, making use of solar input in a regulated manner and avoiding excessive air leakage. It will then link the whole question of thermal insulation to the other contemporary problem – condensation.

Why has condensation suddenly become a fact of everyday life? Is it brought about by the way we live, by the way we construct our buildings, by our ineffective heating or by the cost of fuel, which encourages us to seal ourselves in our unventilated dwellings? Is insulation itself a contributory cause? These questions will be examined and an indication given of how best to avoid the problems.

In addition, the ways in which existing buildings can be treated to improve their thermal performance will be studied, with regard to not only the running costs incurred in producing conditions of thermal comfort but also the minimisation of condensation – one of the major incipient perils of the present day.

REFERENCE

1. Marsh, Paul, *Air and rain penetration of buildings*, Construction Press, 1977.

1 The energy background

For many thousands of years man's society was based on animal and human work-power. The average man's energy output is around 100 watts, with the potential of short bursts of more strenuous effort up to 300 watts. His animal labourers, on the other hand, were capable of higher outputs: the ass 180 W, the ox 500 W and the horse 750 W; consequently man relied upon them to supplement his own work capability, using them for fetching and carrying and working simple machines.

The early attempts to harness natural forces by mechanical devices produced higher outputs than the animals could achieve and dramatically increased man's work power (the water wheel giving 1.5–3.8 kW and windmills 1.5–6.0 kW), but the real breakthrough occurred when man discovered ways of releasing the fossilised sun energy which had been stored, bit by bit, as chemical energy in plant and animal body tissue and then, over millions of years, under favourable geological conditions, converted into coal or oil. The first generation of new inventions unleashed by this discovery – the early steam engine – had energy outputs from 5.2–7.5 kW; succeeding generations of inventions unbelievably augmented these outputs: the present-day 1000cc car produces 45–60 kW, the steam turbine up to 100 MW. Each generation had, in-built, a greater and greater appetite for the fossilised sun power, the magic that started the whole thing off. The energy explosion was soon well under way.

The society man has now created demonstrates an apparently insatiable demand for power. During the last four decades the world's energy production has increased by almost 500% (Figure 1.1) and 97% of today's energy requirement is supplied by fossil fuels – coal, oil and natural gas. We are consuming fuels which, to all intents and purposes, are non-regenerating. Just as we were once in danger of running out of trees to burn, so today we are running out of our newer fuels. A mere 0.02% of the total sun's energy falling on the earth enters the biological system through photosynthesis. It is this energy which, thousands of years later, with a fair amount of good luck and many changes in the earth's crust, may create a new coal seam or oil field. But such long-term promise is in no way going to help man, who is digging the coal and pumping the oil at rates far in excess of even limited replenishment. The fact is this: the stocks of fossil fuels will never be replaced. The available hoard of wealth under the earth's surface is strictly finite.

If we continue to exploit these known, but limited, resources at 1972 rates, we are probably left with sufficient oil for only 30 years and coal for a few hundred years. However, it is highly unlikely that we can keep our apparently ever-increasing demand for power to 1972 levels. Therefore, if an exponential growth of consumption is taken into account, some authorities believe that coal will last for about 110 years and oil for about 20 years.

There are varying forecasts which make allowance for estimated future discoveries of fuel stocks, as well as for the reduced rate of consumption which rising fuel costs, more effective use of fuel and management of fuel resources could bring about. The results of such constraints are already becoming apparent. These forecasts predict that oil could run out between the years 2050 and 2100 and coal between 2500 and 2800. However, all forecasts are to some extent crystal-ball guesses. Only one fact is incontrovertibly true:- fossil fuels are going to run out – and in terms of geological time they will run out tomorrow!

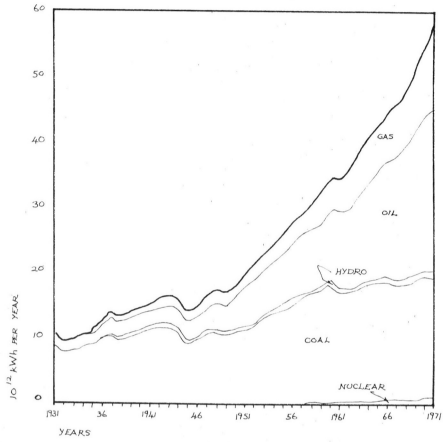

Figure 1.1   World energy production, 1931 to 1971

The even more tragic fact, of course, is that coal and oil are not merely fuels. They are also important raw materials for our chemical industries, raw materials we would find it very difficult to do without. There may come a time when coal and oil may be used only as raw materials in certain designated industrial processes of national and international importance, and their use as fuel will be banned. This is not as far-fetched as it may appear. To burn such valuable materials merely to light and heat our homes or power our motor-cars is clearly nonsensical.

But where is all this energy going at present? In 1972, before the oil crisis, the UK consumed 8.83×10^9 GJ of primary energy, that is, total energy before the losses brought about by conversion of energy from one form to another (for instance the refining of petroleum, or the generation of electricity). Nearly half of this primary energy was from oil (48%), while 37% was derived from coal, 12% from natural gas and 3% from nuclear or hydro power. These statistics represent a dramatic shift of emphasis from coal, which ten years previously had fulfilled 70% of all the UK's energy needs.

It has been stated[1] that buildings – their heating, lighting and the power to run all the equipment they contain – account for 40%–50% of the primary energy consumed by this country. Domestic buildings, which use around 30% of the energy needs of the whole country, require 64% of their total energy consumption for space heating –

nearly 20% of society's total energy consumption for all purposes – and this percentage does not include the requirements for space heating in all other categories of building. From this it can be seen that space heating has one of the largest appetites that the dwindling energy reserves have to satisfy.

Obviously, the hunt for alternative sources of power is on – solar energy, aerogeneration, methods of harnessing the tides, hydro energy in those parts of the world with this potential, and nuclear energy with all its fearful attendant doubts – but our efforts up to the present time have not been very encouraging.

New technologies take time to grow. In the meantime, the absolute necessity is to use our valuable fossil fuel assets as economically as possible.

Since the Industrial Revolution we have been living in a society that has been used to cheap fuel. We have, therefore, often wasted it thoughtlessly. Our comfort norms have risen, and we demand more energy to heat us and light us and run all the machinery of our complex lives. And because it has not cost too much, we have never questioned the validity of it all. But now is the time to question how we live and all those things that have become 'necessities' of our lives.

To start at the beginning; not all fuels are equally efficient in terms of the utilisation of primary energy. Electricity loses more energy, for instance, in its generation than coal or natural gas. Table 1.1 gives a list of ratios of primary energy input to net usable energy produced. These data have led the BRE Working Party[1] to suggest that there should be a switch from electricity to gas and oil for domestic and commercial space heating. This is an example of the type of fundamental re-assessment required. As a rider to this proposition, should not the reject heat from power stations be employed in district heating schemes rather than just being allowed to go to waste as so often happens at present?

Final energy form	Ratio between primary energy input and final energy used
Electricity	3.73
Manufactured fuels	1.40
Oil	1.08
Natural gas	1.06
Coal	1.02

Table 1.1 Primary energy ratios

Other examples of basic questions we should be asking include: Do we need to under-dress, by Victorian standards, to the extent we now do indoors? We have become used to a higher and more even level of indoor heating; but is this completely necessary to our way of life? Alternatively, do we really need all those gadgets that have become a part of our life – and if we do not, maybe the car should go, or the washing machine, or the colour teleivision?

Indeed, all these things need to be questioned; but still we must accept that:-

1. Reasonable comfort levels need to be maintained and, with them whenever possible, those devices which add realistically to the quality of our lives, and
2. We must maintain these standards without endangering the next generation's chances of survival.

If we do not achieve the first, we return to the cave ultimately. If we do not heed the second, we consign our children to the cave with total certainty.

One of the largest consumers of energy, as we have seen above, are buildings – and particularly the heating of buildings. Therefore, the primary obligation of every designer must be to produce thermally efficient buildings. The BRE Working Party suggested that a 15% reduction in total primary energy consumption in the UK could be made in all buildings – both new and existing. This is not an insignificant proportion. Bearing in mind the huge stock of existing buildings which can be only superficially treated, the contribution of new buildings must be as effective as possible. This is not just a matter of thermal insulation in the external elements of the shell, although this is a very important factor; it is a matter of designing from the start with the primary object of producing a building that not only satisfies the practical, day-to-day, functional requirements of that building, but does so in a manner that produces a thermally efficient structure.

The running costs of the space heating of buildings are already not cheap – gone forever are the heady days of inexpensive fuel – but before long they will become astronomic. Any building designer who is looking after his client's best interests should have thermal efficiency near the top of his list of design obligations. Without the building being correctly related to its site, thoughtfully fenestrated and orientated and constructed of thermally efficient enclosing elements, the benefit gained from more effective heating appliances is negated.

Thermal design is a team operation with the services engineer an indispensable member of the design team, almost from day one. The architect can no longer risk designing his building and then presenting it to the services engineer to heat and cool as may be necessary. The design process is a dialogue between inter-dependent skilled professions. Just as the architect and structural engineer should be involved in a continuous exchange of ideas, so should the architect and the services engineer.

This will place pressure on the services engineer to take up a rather different role from that for which he has been trained. He has been equipped to produce heating and cooling systems for any form of building in any type of climate, using a series of accepted criteria, because that was all he has previously been expected to do. Now the requirement must be for him to influence the architect's design decisions by his particular expertise. In other words he is required to provide a positive design input in order to achieve a building which will provide its occupants with thermal comfort at a reasonable cost.

The following chapters outline some of the ways in which the design and siting of a building can aid the achievement of this objective, providing the architect not with the complete solution, but with a framework of knowledge and a language with which to discuss the problem effectively with the services engineer.

REFERENCE

1. BRE Working Party, *CP56/75, Energy conservation: a study of energy consumption in buildings and possible means of saving energy in housing*, BRE, 1975.

2 Thermal comfort – the physiological background

The sensation of thermal comfort, as with all physiological sensations, is notoriously subjective. Thermal comfort has been defined as that condition in which over 50% of the people are unaware of their thermal environment; that is, they feel no need to adjust it. For only just over 50% of the people to be entirely satisfied (or, put another way, to be totally unaware of their thermal surroundings) suggests a tremendous range of individual response – and this is precisely the case. The feeling of comfort varies from person to person, depending on age, sex, nationality, habit and type of activity being undertaken. It is also subject to diurnal and seasonal acclimatisation. Our tolerance of temperature variation can be modified by use.

Over the last twenty or thirty years this tolerance level has dramatically fallen, placing more stringent demands not only on the methods of heating interior spaces but also on the thermal performance of building shells. Our grandparents huddled round the open fire, their knees scorching, their backs frozen. They expected temperature variation in their buildings, not merely between one room and another, but even between different positions in the same room. Physiologically, they were acclimatised to this temperature unevenness; practically, they wrapped themselves in rather more layers of clothing than is our present habit.

In fact, we have grown used to living in warmer and more evenly heated surroundings. As a result of this, we have shed some of the heavy clothes which our grandparents accepted as normal attire and replaced them with lighter, less cumbersome clothing.

The purpose of clothes is to insulate the body – to reduce its heat loss to the surrounding environment. The body produces heat, some of which, unless over-heating and discomfort are to occur, needs to be dissipated to the surroundings. It is usually found that when the mean skin temperature is maintained between 31°C and 34°C, the majority of persons feel comfortable. If the skin temperature is automatically maintained at this level, thermal equilibrium exists, which means that there is a physical balance between the temperature surrounding the body, the body itself with its layers of insulation and the activity being undertaken. The BRE have christened this condition a state of thermal neutrality.

A scale of insulation values for clothing has been devised employing a unit called a Clo. One Clo is roughly equivalent to a normal suit (a thermal resistance approximately equal to 0.155 m²°C/W), ½ Clo is equivalent to shirt and trousers, or blouse and skirt (typical indoor summer wear) and so on.

The average British person's daily energy expenditure (excluding the more active occupations) is 67 W/m². Sedentary tasks produce a lower metabolic rate, light housework a slightly higher one (about 78 W/m²) and moderate housework about 110 W/m². To produce a state of thermal neutrality, clothing, activity and room temperature have to be balanced and this is shown graphically in Figure 2.1.

From this it can be seen that a comfortable temperature for sedentary occupations with a clothing level of 1 Clo would be 20°C.

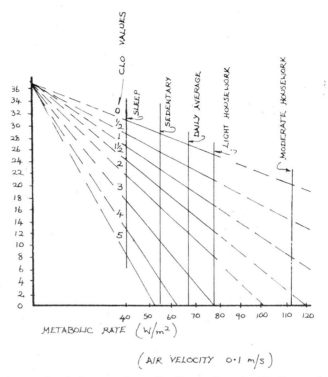

Figure 2.1 Approximate temperatures for thermal neutrality for various weights of clothing

It seems certain that the average clothing for indoor wear fifty years ago would be at least $1\frac{1}{2}$ Clo. This amount of clothing we would find inconvenient and uncomfortable today, although it is not beyond reasonable expectation that clothing with similar (or better) thermal insulation qualities could be developed, which would still give freedom of movement in wear and be wholly uncumbersome. An extra $\frac{1}{2}$ Clo would reduce indoor comfort temperatures by 3°C and make a dramatic 30% reduction in the energy required for heating – in times of energy scarcity, a significant saving.

The establishment of comfort levels is a complex procedure. The results of examining the reactions of primary school children to their thermal surroundings[1] have highlighted how subjective the whole problem is. For instance, it was found that children who were more heavily clad did not on average report feeling warmer; also, all children tended to react more positively to a temperature change rather than to a steady low temperature; both results indicating the principle of acclimatisation. A further interesting fact was that the children tended to adjust their clothing thickness in response to long term temperature changes, but failed, generally, to make short term adjustments to alleviate diurnal variations. They demonstrated a rather greater range of tolerance to steady temperatures than might have been expected – the range of temperature which apparently they found acceptable being from 17°C to 23°C. This tolerance could, however, have resulted from varied activity levels.

The problem of thermal comfort is clearly complicated by the fact that people differ in their reactions to indoor thermal environments. For instance, there is approximately 15% variation in the daily energy expenditure of individuals leading apparently similar lives. This can produce a deviation of 2°C, if reflected in indoor temperature levels. Also, individual response to room temperature can produce deviations of 2°C to 3°C. An individual is not always consistent in his sensations of warmth on different occasions in the same room at the same temperature. Additionally, between the ages of 25 and 80 there is a gradual reduction of about

15% in the energy expenditure of a seated person. Therefore, the older one becomes the higher the room temperature has to be for one to experience the same degree of thermal comfort.

However subjective the sensation of thermal comfort may be, it is still possible to establish basic recommendations which will provide a thermal environment acceptable to the majority of people, clad in an average manner or, at least, in a manner to which they are accustomed. As a rule, within each recommended temperature range, the lower temperatures will be more acceptable to the young male, the upper temperatures to the more elderly and the majority of females.

Thermal comfort of the individual is not merely a matter of heating. We have already seen that it is influenced by activity levels and clothing. In addition, it depends on the thermal characteristics of the building enclosing the individual, levels of humidity within the occupied space and the amount of air movement. All these factors affect the exchange of heat between a person's body and its surroundings, and the rate of this exchange produces sensations of warmth or coldness.

The majority of the energy derived from a person's food is emitted in the form of heat. Normal activities liberate approximately 110 W, 80% of which is emitted by convection and radiation and 20% by evaporation (mainly from the lungs). Table 2.1 gives an indication of how the quantity of heat emitted varies with the type of activity – the more strenuous the activity, the more heat is produced. As the body requires to maintain its own temperature at a relatively stable level, the ease or otherwise with which it disposes of or retains this heat is a vital factor in general bodily comfort. The body has its own involuntary mechanisms for adjusting its temperature when the conditions of its surroundings are not such that this adjustment of heat can take place automatically. For instance, when the total heat loss of a body is insufficient to avoid a rise in body temperature, the body sweats, thereby aiding cooling (or heat loss) by evaporation. If the surrounding humidity is high (a common situation in many tropical areas), evaporation is hindered and the body consequently suffers discomfort. When body heat loss is too rapid (experienced in under-heated rooms used for sedentary activities) the blood supply to the extremities is reduced and the muscles endeavour to produce the additional heat required to compensate for this loss by shivering. Obviously, a thermally comfortable environment is one in which such involuntary body mechanisms are unnecessary.

Activity	Total heat production by the adult male (W)
Immobile – reading	110–115
Light work	140
Walking slowly	160
Light bench work	235
Medium work	265
Dancing	265
Heavy work	440

Table 2.1 Heat output related to activity.

From this it can be seen that thermal comfort depends on a number of qualities of the indoor space. These can be listed as:

1. Air temperature.
2. Surface temperature.
3. Air movement.
4. Humidity.

The establishment of some method of measurement of comfort involves an assessment of all four qualities and is, consequently, a complicated process. In effect, what one is endeavouring to do is to arrive at a measure of a person's *feeling of warmth*, which need not necessarily be reflected truthfully in measurements obtained by merely rotating a thermometer in the air.

We must all remember having said, at some time, to a friend when we met in the street, 'Come round the corner out of the wind. It'll be warmer there.' It will not be warmer. The air temperature will remain unchanged. Heat loss from the body, however, will be less out of the wind and there will be a *sensation* of greater warmth.

Heat loss by convection is dependent on the temperature and humidity of the surrounding air, as well as the rate at which the air makes contact with the exposed skin and, to a lesser extent, clothing. The rate of radiant heat loss is governed by the difference in temperature between the exposed skin and clothing and the temperature of surrounding surfaces. The most dramatic illustration of this is the discomfort experienced when one is sitting indoors next to an uncurtained window when the external temperature is low. This experience need have nothing to do with draughts from the window, or cold downward air movement adjacent to the glass. It is purely a matter of radiant heat loss to an adjacent cold surface.

THE MEASURING OF THERMAL COMFORT

For the last fifty years efforts have been made to establish a reliable method of measuring thermal comfort – in effect, to define a *comfort temperature*. In the 1920s Houghten, Yaglou and Miller devised a subjective scale of assessment based on a series of people's reactions when moving from one environment to another. This they called the *effective temperature* scale. It was modified in 1932 by Vernon and Ward to take account of radiant heat. This *corrected effective temperature* scale was still a subjective method, now shown to over-estimate in certain respects and, therefore, no longer recommended for use.

Equivalent temperature   Equivalent temperature, which was recommended in BS CP 3: Chapter VlII: 1949, makes an attempt to combine the effects of air temperature and radiant heat from room surfaces by a non-subjective method. It is defined as 'that temperature of a uniform enclosure in which, in still air, a black cylinder, of height about 22 in and diameter about $7\frac{1}{2}$ in, would lose heat at the same rate as in the environment under consideration, the surface of the cylinder being maintained at a temperature which is a precise function of the heat loss from the cylinder, and which in any uniform enclosure is lower than 100°F by two-thirds of the difference between 100°F and the temperature of that enclosure.' Equivalent temperature is measured by a eupatheoscope and is derived from the following equation:

$$t_{eq} = 0.522t_a + 0.478t_r - 0.21(37.8 - t_a)\sqrt{v}$$

in which t_{eq} = equivalent temperature (°C)

   t_a = air temperature (°C)

   t_r = mean radiant temperature of surrounding surfaces (°C)

   v = air velocity (m/s)

This scale takes no account of humidity and is, therefore, inappropriate at high temperatures when humidity becomes a significant factor in the ease with which the body can lose its heat. Equivalent temperature has tended to fall into disuse in favour of global temperature and dry resultant temperature.

Global temperature   Global temperature is explained in the CIBS Guide Book (previously the IHVE Guide Book) and is similar in effect to equivalent temperature. It correlates well with subjective assessments of warmth, amalgamating all aspects affecting comfort except, once again, humidity. It, therefore, suffers from the same shortcomings as equivalent temperature. It is measured by a thermometer in a 150 mm diameter blackened globe. It can be calculated from the equation:

$$t_g = \frac{t_r + 2.35t_a\sqrt{v}}{1 + 2.35\sqrt{v}}$$

in which t_g = global temperature, all other symbols being as in the previous equation.

Dry resultant temperature   Dry resultant temperature is becoming more widely used and is, in fact, the recommended method for assessing comfort temperatures in the CIBS Guide. The method has great popularity in Europe. It was devised by A. Missenard in 1935 and, for this reason, is sometimes measured in °M (degrees Missenard) on the Continent in honour of its inventor. Like global temperature it is measured by a thermometer inside a blackened globe – in this case a 100 mm diameter globe. Recent results of BRE research[2] suggest that an even smaller globe (40 mm dia) is more convenient to use, quicker in reaction and preferable when used as an idex of subjective warmth. Dry resultant temperature is a little less sensitive to radiant temperature than previous methods and it still takes no account of humidity. Dry resultant temperature can be derived from the following equation:

$$t_{res} = \frac{t_r + 3.17t_a\sqrt{v}}{1 + 3.17v}$$

When there is no substantial air movement t_{res} is approximately $\frac{1}{2}t_r + \frac{1}{2}t_a$. In the revised CIBS Guide (at the time of writing it is in preparation) Part A1 establishes the dry resultant temperature as the *comfort temperature* and the one to be used in design calculations.

There is also a *wet resultant temperature* which takes into consideration humidity.

Of the three main methods of establishing thermal comfort, ease of measurement favours either global or resultant temperature, rather than equivalent temperature. All give reasonable assessments of comfort when air speeds are low and when radiant and air temperatures do not differ greatly and humidity is not significant. It is for this reason that the CIBS Guide recommends the dry resultant temperature as the comfort temperature. In previous editions of the Guide (particularly IHVE Guide 1970) *environmental temperature* was used as a simplified equivalent of comfort temperature. It was derived from the formula:

$$t_{ei} = {}^2/_3t_r + {}^1/_3t_a$$

This is now superseded by the simplified version of dry resultant temperature:

$$t_{res} = \tfrac{1}{2}t_r + \tfrac{1}{2}t_a$$

It is a close approximation to the actual comfort state and will be referred to throughout this book as t_c.

It should be remembered that the approximation in the equation above is only valid when the air movement is small. In other conditions the t_{res} (t_c) should be calculated using the full formula given earlier.

CRITERIA OF THERMAL COMFORT

These recommendations can be considered as being generally acceptable in Britain, the rest of Europe and America.

Sedentary occupations   A seated person expends 50 W/m² of energy when totally passive (reading or watching television) and 60 W/m² when indulging in low level activity, such as clerical work or sewing. Taking the average value of 55 W/m² and assuming normal winter clothing of 1 Clo, 23°C appears to be the temperature of comfort. This is at the top of the usually accepted range of resultant temperature for areas of sedentary occupations (19°C–23°C), probably due to the fact that the insulation provided by the chair on which the subject is sitting has been ignored in the assessment. The normal range of 19°C to 23°C is, therefore, recommended, but this assumes certain other temperature distributions.

Ideally the mean radiant surface temperature should be slightly higher than air temperature. Certainly the mean radiant temperature should not vary from the air temperature by more than 8°C higher, or 3°C lower. If stuffiness is to be avoided, the air temperature should never be allowed to be considerably higher than the mean radiant temperature.

The temperature gradient from floor to ceiling in low rooms should not be greater than 5°C, while the maximum differential between floor and head height should be 3°C. Preferably the temperature at floor level should be slightly higher than at head height. The widest temperature gradients (and, therefore, those potentially liable to produce discomfort) occur in rooms heated by convection; the smallest in rooms heated by radiation.

In sedentary areas relative humidity normally, in temperate climates, has little effect on feelings of comfort. Very high humidities, however, will accentuate feelings of chilliness at low temperatures and cause oppressiveness at high temperatures. Very low humidities can cause dryness in the throat. Generally, relative humidities should be maintained between 40% and 70% (BS CP 3; Chapter 11; 1970 recommends between 40% and 80%).

Figure 2.2 indicates maximum recommended air movement velocities, in relation to the temperature of the moving air. This is reproduced from DIN 1946. As can be seen, the warmer the air, the greater can be its velocity without causing discomfort. The dotted curve B indicates more acceptable conditions when the air movement is directed at particularly sensitive parts of the body, such as the back of the neck. All speeds in this graph should be considered maxima and lower speeds should be achieved wherever possible.

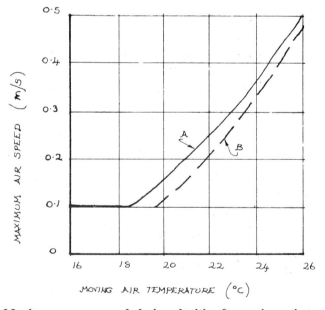

Figure 2.2 Maximum recommended air velocities for various air temperatures

The energy expended when sleeping or resting is in the region of 40 W/m². A person in bed loses relatively little heat downwards through the mattress of the bed, therefore the effective heat flow from the body is only about ²/₃rds that of a person standing or sitting in an upright chair. Blankets having an inherent insulation of 1 Clo will therefore have an *effective* insulation of 1.5 Clo. The minimum amount of bedding used is probably one sheet and one blanket, which could be hardly less than 1 Clo. Applying this 1 Clo insulation to a metabolic rate of 40 W/m² the maximum room temperature should be 27°C. This is obviously excessive, as a sleeping person with reasonable bedding can be comfortable with a room temperature of 0°C. He would, however, complain during dressing and undressing, or while reading in bed.

The lower limit of temperature in a bedroom should therefore be 15°C, which produces a requirement for bedding having the equivalent of 2.5 Clo units – or two good blankets and a sheet.

In spaces where people are likely to be unclothed or partly clothed, such as bath halls, slightly warmer conditions with slower air speeds than are normally acceptable in sedentary areas are recommended. Resultant temperatures between 25°C and 30°C should be provided, with only a small difference between air temperature and mean radiant temperatures and with air speeds not exceeding 0.1 m/s at 25°C.

Active occupations

There has been relatively little study made of environments in which heavy physical work is to be performed. Work habits tend to complicate the picture. During hard work it is likely that some clothing will be discarded, both for thermal comfort and for freedom of movement. In addition, energetic work tends to be performed in short bursts, punctuated by rest periods. Both these habits reduce the amount by which the resultant temperature can be lowered. Generally, a reduction of 3°C to 5°C appears acceptable. The higher of these values would be appropriate for activities producing 100 W of body heat in excess of normal sedentary output.

At higher levels of activity, particularly in warm environments, sweating is a normal route for body heat loss and, therefore, the relative humidity should be maintained at a fairly low level; though even in extreme cases a relative humidity below 40% would hardly be comfortable.

In designing work spaces, the architect should, of course, bear in mind the requirements of the Factory Acts regarding thermal environment.

If the resultant temperature is appropriate to the work level, similar criteria apply to air movement relative to air temperature in active and in sedentary areas. If, however, the environment is too hot, higher air speeds are permissible, because they lead to increased body heat loss through evaporation.

Summer conditions

It is interesting to note that in summer there is a tendency for a person's thermal comfort range to rise by 1°C or 2°C. This is probably due both to acclimatisation and to the wearing of lighter clothes. Sedentary areas, therefore, can be allowed to increase to between 20°C and 25°C and working areas to between 15°C and 22°C.

Ventilation

Another ingredient of the comfortable environment, which has implications in the design of the building envelope and its thermal performance, is ventilation. The introduction of fresh air and the removal of vitiated air is an obvious necessity of any occupied indoor space. Not only does it provide oxygen for breathing and removes stale, contaminated air; it also removes water vapour (an aspect of importance in relation to condensation, which is considered later in this book) It also results in heat loss. Whether ventilation is achieved by natural means (openable windows) or by mechanical extraction or air-conditioning, allowance needs to be made in the sizing of the building's heating system for this heat loss by ventilation.

The designer of the building shell can assist the heating engineer by avoiding, as far as possible, fortuitous air leakage through his structure, which can upset the heating levels or the anticipated air movements. An allowance is made in heating designs for specific levels of fortuitous leakage, but these levels must not be exceeded by failures of the building shell to perform efficiently, or by a wrongly selected window type for a particularly exposed location (see Chapter 4).

Table 2.2 gives a list of various indoor spaces and the mandatory, or recommended, environmental standards which are to be achieved to provide thermal comfort.

THE EFFECT OF THE BUILDING SHELL ON THERMAL COMFORT

The shell of a building has a major influence on the attainment of conditions of thermal comfort.

Type of Building	t_c (°C)	Air infiltration rate (h⁻¹)	Ventilation allowance (W/m²°C)
Art gallery, museum	20	1	0.33
Assembly halls	18	$\frac{1}{2}$	0.17
Banking halls	20	1	0.33
Bars	18	1	0.33
Canteens, dining rooms	20	1	0.33
Churches and chapels	18	$\frac{1}{2}$	0.17
Dining, banqueting halls	21	$\frac{1}{2}$	0.17
Dwellings, hostels:			
Living rooms	21	1	0.33
Bedrooms	18	$\frac{1}{2}$	0.17
Bed-sitting rooms	21	1	0.33
Bathrooms	22	2	0.67
Lavatories, cloakrooms	18	$1\frac{1}{2}$	0.50
Exhibition halls	18	$\frac{1}{2}$	0.17
Factories:			
Sedentary work	19	5–6 litres per person if no smoking, 12 litres if smoking allowed, or light and heavy work.	
Light work	16		
Heavy work	13		
Gymnasia	16	$\frac{3}{4}$	0.25
Hospitals:			
Corridors	16	1	0.33
Offices	20	1	0.33
Operating theatres	18–21	$\frac{1}{2}$	0.17
Stores	15	$\frac{1}{2}$	0.17
Wards and patient areas	18	2	0.67
Waiting rooms	18	1	0.33
Hotels:			
Bedrooms (standard)	22	1	0.33
Bedrooms (luxury)	24	1	0.33
Public rooms	21	1	0.33
Corridors	18	$1\frac{1}{2}$	0.50
Foyers	18	$1\frac{1}{2}$	0.50
Laboratories	20	1	0.33
Libraries:			
Reading rooms	20	$\frac{1}{2}$	0.17
Stack rooms	18	$\frac{1}{2}$	0.17
Store rooms	15	$\frac{1}{4}$	0.08
Offices:			
General	20	1	0.33
Private	20	1	0.33
Stores	15	$\frac{1}{2}$	0.17
Restaurants	18	1	0.33
Schools and colleges:			
Classrooms	18	2	0.67
Lecture rooms	18	1	0.33
Shops and showrooms:			
Small	18	1	0.33
Large	18	$\frac{1}{2}$	0.17
Department store	18	$\frac{1}{4}$	0.08
Fitting rooms	21	$1\frac{1}{2}$	0.50
Store rooms	15	$\frac{1}{2}$	0.17
Sports pavillions:			
Dressing rooms	21	1	0.33
Swimming baths:			
Changing rooms	22–24	$\frac{1}{2}$	0.17
Bath halls	26–28	$\frac{1}{2}$	0.17
Warehouses:			
Working and packing spaces	16	$\frac{1}{2}$	0.17
Storage spaces	13	$\frac{1}{4}$	0.08

Table 2.2 Environmental standards for comfort.

In effect, the building shell makes the first, coarse adjustments to the external climate, producing an enclosure with an environment which can then be finely tuned to acceptable levels of comfort by heating, cooling, ventilation and air-conditioning. Afterwards it helps to preserve this unnatural environment by its 'overcoat' effect – its quality of thermal insulation and thermal response. The more efficiently the building shell functions in these respects, the more effective and economic will be the heating installation, and the greater the saving in fuel.

In the past many of these design qualities have been automatically contained in vernacular building – part of the tradition which was passed down from generation to generation of craftsmen; men who were sometimes consciously aware of what they were doing, though often largely unconscious of all the implications of the way they built.

Instances of this can be seen in the siting of farm buildings in sheltered valleys, the avoidance of sites on windward slopes, the positioning of outbuildings to protect the living quarters, the use of high thermal capacity structures with small windows – usually on the less exposed elevations. All these design habits have become irrelevant, or uneconomic, or simply impossible in these days of expensive building, over-competition for diminishing land resources and the consequent high land costs. In addition, the prodigious technological flowering of the last century has produced different building needs and new materials of construction, bringing changes in the design vocabulary.

In the flurry of new design thought, there was often a failure to analyse why building had been carried out as it had in the past. There developed the cult of the new and a rejection of the old. Many lessons which our great-great grandparents had learnt have been lost – and now are being painfully learnt again from basic principles. Designing for an efficient, economic-to-run internal thermal environment was one of these. This, together with the reduction in tolerance levels referred to earlier, the no longer cheap fuel and changing ways of life – all these have led to a proliferation of thermally wasteful structures. Waste often became synonymous with the new design vocabulary. The excessive use of glass was just one example.

In order to avoid the production of buildings with poor thermal performance it is not enough to insulate the exterior elements (although this is clearly important). The building should be conceived with thermal efficiency as one of the objectives of the design, equal in importance to other objectives of function, structural soundness or aesthetics.

In conclusion, the primary function of a building shell is to develop the best and to ameliorate the worst of the climate indoors and out-of-doors. To this end it should exclude the less desirable elements – rain and snow – while allowing the more desirable parts of the climate to penetrate in a controlled manner – daylight, sunlight (subject to the avoidance of excessive solar heat gain) and fresh air in naturally ventilated buildings, while still avoiding draughts and excessive heat loss.

The contribution of the building shell to the thermal comfort of its occupants is paramount, and it makes its contribution in the following ways:

1. In its design and layout, its relationship to the site, and its mass – size, shape and weight.
2. In its ability to exclude or control the potentially undesirable parts of the external climate – air filtration and excessive solar heat gain.
3. In the thermal performance of its enclosing elements which preserve a consistent internal thermal environment in spite of a considerable difference between that and the external climate.

Each one of these aspects will be examined in succeeding chapters.

REFERENCES

1. Humphreys, MA, *A study of the thermal comfort of primary school children in summer*, Current paper CP17/78, BRE.
2. Humphreys, MA, *The optimum diameter for a globe thermometer for use indoors*, Current paper CP9/78, BRE.

3   Criteria in the design and siting of the building shell

Right at the outset, almost before pencil is set to paper, broad design principles neea to be considered by the architect, if he is to conceive a thermally efficient building. Before he gets engrossed in what his brain-child will look like, or how it will be constructed, or even how to insulate its walls and roof, he needs to consider the setting of the eventual building in the micro-climate which exists on its site. What special environmental hazards does this particular site hold? How best can its climatic advantages be exploited? What sort of building, deployed upon the site in what way, will prove most thermally effective? He needs to consider the external mass of his building – its size and shape and thermal 'weight' – as a heat containing (or, in hot climates, excluding) envelope. Will it be, from a purely geometric standpoint, thermally efficient? As the design progresses he needs to ask, Will it make the best use of the energy that will be pumped into it during its lifetime? Will it preserve that energy and use it prudently? Then the architect needs to look a little more closely at the inside of the building and assess the way in which it contributes to the preservation of its own internal climate.

RELATIONSHIP TO THE SITE

Exposure

Rarely will the building designer be able to influence the selection of the site for the building which he has been commissioned to design. Usually he is presented with a *fait accompli* and discovers that he is expected to design a thermal miracle of a building on a cliff top overlooking the Atlantic, or on a south-west slope of the Pennine hills.

Clearly, the more sheltered the site, the easier (and therefore the less expensive) it will be to exclude the elements. Of these, the wind can produce unwanted air infiltration which can upset the designed number of air changes in a naturally ventilated building, thereby placing an unexpected load on the heating plant, or upset the balance of the air-conditioning equipment of a more sophisticated building (see Chapter 4). The slope of the site and the general terrain surrounding the site, the relationship of that slope to the prevailing weather quarter and to the sun path – all will influence the micro-climate of the site. The thoughtful designer should consider this when developing his building design. If he is initially involved in the *choice* of the site, such characteristics must influence his decision.

The factors most likely to have influenced the selection of the site by others would probably be:

1.   Accessibility for people and for the delivery of goods.
2.   Proximity requirements of the occupants – closeness to clients, customers, raw materials, suppliers, shops, places of work, roads, railways, docks etc.
3.   Availability of sites at an economic price.
4.   The prestige value of the site.

The climate of the site would not normally have figured in the decision-making process. The choice could have been made as the result of extensive market research, an instant, expedient, committee decision, or economic pressure. The designer has to make the best of the site he is given.

The exposure of the site and its surrounding areas, the presence of natural or artificial obstructions to the wind, and the height of the proposed building in relation to its surroundings will all have their effect on the rate of heat exchange between the building's surfaces and the external air. Additionally, they will affect the rate of air infiltration.

BS CP 3: Chapter 11: 1970 sets down a simple scale of exposure classifications and the mean wind speeds likely to be experienced in each:

1. Sheltered: the first two storeys of buildings in towns or built-up areas; likely wind speed 1.0 m/s.
2. Normal: suburban houses and the third and fourth storeys of buildings in towns; likely wind speed 3.0 m/s.
3. Exposed: unscreened buildings in inland areas and for storeys above the fourth in towns; likely wind speed 9.0 m/s.

This table of exposures should not be confused with a similar one, quoted in Chapter 4, which is derived from the CIBS Guide 1970 for establishing exposure grades for windows. In fact, the whole subject of exposures measurement is becoming very complex in this country at the present time. In questions of rain penetration, for instance, the driving-rain index and not the wind speed is used as a measure of exposure; and yet in a draft British Standard covering window performance (including rain leakage), exposure is defined by reference to three-second gust speeds. (This draft BS is also referred to in Chapter 4). To add a further complication, wind speeds are not always what they seem. The three-second gust speed is used in the draft BS dealing with window performance – part of which deals with air leakage, which is a component of natural ventilation – and yet BRE Digest 210 on natural ventilation uses hourly mean wind speeds. It is all rather confusing and the best advice is always to establish precisely what definition of exposure and wind speed is being referred to on all occasions.

Returning to the table of exposures quoted from BS CP 3: Chapter 11: 1970, this gives a very simple means of assessing the mean wind speed applicable to the majority of UK sites. The wind *speed* so obtained can then be applied to the formula given in Chapter 4, page 30, to establish a wind *pressure*, from which air infiltration amounts can be calculated. The CP list is, however, limited. It does not cover positions of extreme exposure, such as coastal or hill crest sites. In these cases, and in others where more accurate information is thought necessary, the method set out in BRE Digest 210, for assessing wind speed for a particular building height on a particular site, should be followed.

In this method, first the area wind speed needs to be known. The Meteorological Office can provide hourly mean wind speeds for most areas; alternatively, reference can be made to maps of hourly mean wind speeds. These wind speeds (U_m) are measured at a height of 10 m in open country and, therefore, need adjustment if they are to be applied to other types of terrain and other heights of building. This can be carried out by using the appropriate conversion factors from Table 3.1 in the following formula:

$$U = U_m \times K_z{}^a$$

in which U = particular mean wind speed (m/s)

U_m = meteorological wind speed (hourly mean) measured in open country at a height of 10 m (m/s)

z = height of building (m)

K = coefficient relating wind speed to height (from Table 3.1)

a = exponent relating wind speed to height allowing for terrain (from Table 3.1).

Terrain	K	a
Open flat country	0.68	0.17
Country with scattered windbreaks	0.52	0.20
Urban	0.40	0.25
City	0.31	0.33

Table 3.1   Factors for determining mean wind speed from wind speed U_m measured as 10 m in open country.

BRE Digest 210 gives a map showing the cumulative frequency of wind speed which will be exceeded 50% of the time (Figure 3.1). These wind speeds (U_{50}) can be modified to take account of greater or lesser frequency of occurrence by the ratios in Table 3.2.

The effect of exposure on surface heat transfer is considered in Chapter 5. From this it will be seen that the external surface resistance coefficient varies in response to the degree of exposure of the site. External surface resistance is, however, only a small percentage of the total thermal resistance of the average structure and any modification of this factor, therefore, makes little difference to the structure's thermal insulation. Only in poorly insulated buildings does a change in surface resistance become significant.

	Location	
Percentage	Exposed coastal	Sheltered inland
80	0.56	0.46
75	0.64	0.56
70	0.71	0.65
60	0.86	0.83
50	1.00	1.00
40	1.15	1.18
30	1.33	1.39
25	1.33	1.39
20	1.42	1.51
15	1.70	1.80
10	1.84	2.03

Table 3.2   Values of the ratio of mean wind speed exceeded for a given percentage of time to the 50% mean wind speed U_{50}.

Nevertheless, however small the effect may be, it should not be ignored when considering the layout of a building in relation to its site. Advantage should be taken of any shelter available. Even though the subsequent improvement in the calculated heat losses may be such as to have no effect on the size of the boiler plant required and, as a result, make no reduction possible in the capital cost of the building, over a long period of use the effects measured in reduced expenditure on heating could be considerable. This is the type of improvement the building designer may be able to make without incurring any increase in the building cost.

The height of a building, too, affects both the heat transfer from its surface and its air infiltration levels. The higher the building, the more exposed its upper floors become. It is worthwhile considering seriously if, say, a seven storey building is really the best solution when the same accommodation could be rearranged in four storeys.

Such decisions as what height the building should be and how it should be deployed on the site, therefore, should be influenced by the nature of the site and its micro-climate. The designer needs to become 'shelter-conscious'. In some cases it

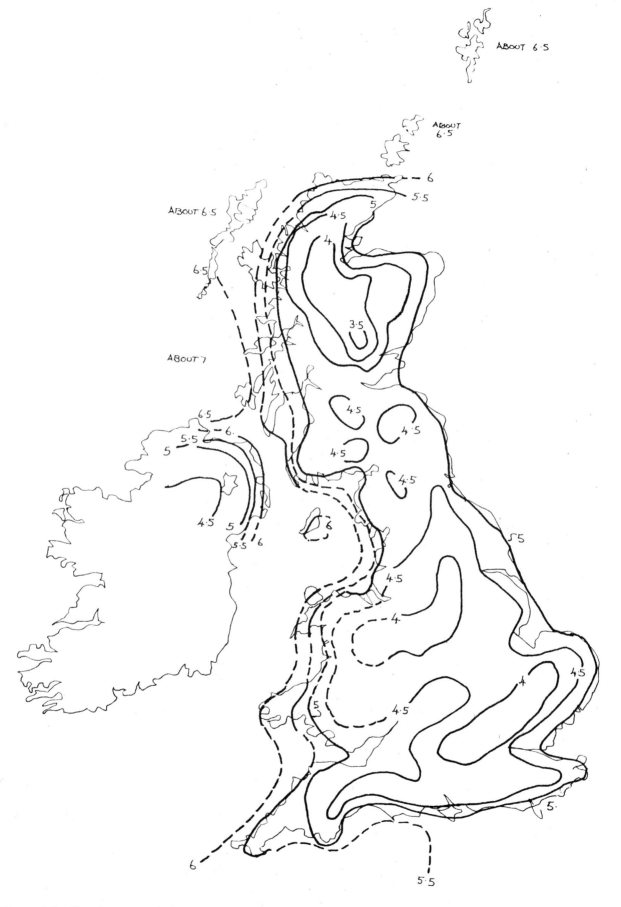

Figure 3.1   Hourly mean wind speed (ms⁻¹) exceeded for 50% of the time

Relationship to the site

might be advisable to seek local meteorological data – certainly to establish from which quarter the worst weather is likely to come. The south-west is not invariably the wettest and windiest compass bearing. Such information will prove useful, in any case, to establish the types of window to be chosen – to keep the water out, as well as the air – and the form of detailing of the building surfaces to avoid the danger of rain penetration.

In trying to make the best use of the site from a thermal point of view, the designer will have to harmonize thermal influences on his design not only with the proper functioning of the building and with occupational requirements which may influence the shape of the building, but also with such other external influences as:

Planning requirements and other statutory obligations, or covenants.
The exploitation of natural amenities (views etc).
The minimising of local nuisances (noise from motorways, railway lines, or neighbouring sites etc).
The avoidance of excessive solar heat gain (Chapter 4).
The load-bearing characteristics of the ground.
The site levels.

Compatibility needs to be sought and a compromise achieved, but the thermal component of this puzzle is a very important one, and will become more important year by year.

Orientation

In addition to any other external influences on the way a building may be deployed on a site, there are solar influences which the designer ignores at his peril. These will be discussed in more detail in the next chapter, but here the principle of orientation and its beneficial possibilities are explained.

Just as fresh air can be stimulating on a spring morning, but a pain in the neck as a winter draught, so the sun can be either a benefit, or an insufferable nuisance. A building needs to represent a delicate balance between total exclusion and total acceptance of the outside climate – and the establishment of this balance begins with the orientation of the building.

The least wasteful building derives from the controlled use of all the benefits of the natural climate and the exclusion of all its less advantageous traits. To accomplish this, the building must be one that absorbs the advantageous input of the natural surroundings, conserving useful energy for immediate, or future, use. This is the *passive* approach, as opposed to the *active* one, which endeavours in creating a building, no matter what its surroundings, to make it so invulnerable that the artificial climate which is pumped into it, by expensive, over-sized and technologically complex mechanical plant, is preserved in spite of the external climate.

The *active* building is largely intractable, creating protection for protection's sake, even when the outside climate is congenial or has congenial components. It is, therefore, a wasteful use of energy, at times using valuable fuel to create comfort which Nature would have provided free of charge given an open window.

The *passive* building, by contrast, is made a willing receptacle within which to capture and preserve natural energy. It is sympathetic to its surroundings and in its orientation, taking advantage of south-facing windows in northern latitudes to acquire solar heat in winter, while avoiding the low-level radiation from the east and west in summer, which would create intolerable conditions without air-conditioning. The building conserves its internal, unnatural climate, without rejecting the more useful components of the natural, external climate.

A good example of the passive approach to thermal comfort is the Autarkic Housing Project in Cambridge (Figure 3.2). Here a dual-zoned house has been designed, providing a variable floor area the use of which will be dependent on the weather conditions. The *closed* zone, to be used in severe weather, consists of the

19

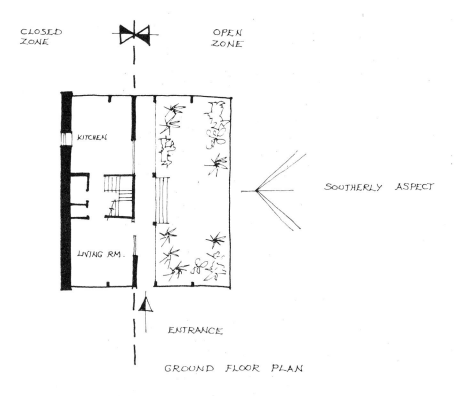

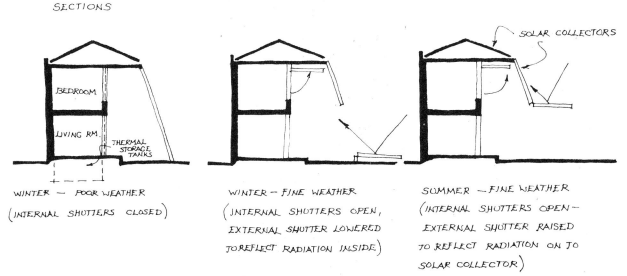

Figure 3.2 The Autarkic Housing Project

normal rooms one would expect to find in an average dwelling – living room, kitchen, bathroom and two bedrooms – 71 m² in area and on two floors, each of 2 m ceiling height. These rooms open (through insulated shutters) on to the *open* zone of the building – a 5 m high enclosed garden, or conservatory with large glazed areas which, under favourable conditions, make the most of direct solar radiation. The *closed* zone can share these conditions at will, merely by the opening of the shutters. It is interesting to note that during an average year there seem to be less than 60 days when this additional area would not be used; but, as far as heating costs are concerned, this area is outside the house.

Orientation will be discussed in relation to solar heat gain in the next chapter, nevertheless here it is important to remember that a stable internal environment is more likely if large areas of glass are avoided on east and west elevations. Glazed areas on south-facing elevations, on the other hand, will help to reduce winter heating costs, so long as the heating system is sufficiently controllable to take account of the temperature differential between north and south sides of the building. Northerly windows will not suffer solar heat gain; they could cause excessive heat loss in winter if they are too large. The optimum orientation of a rectangular block in northerly latitudes appears to be with its long axis running east and west. Its south-facing glazed areas should be equipped with some form of sun-shading devices and its north-facing glazed areas should be not too extensive in area. In southern latitudes the orientation of the block would remain the same, but the comments on glazed areas would be reversed.

BUILDING SHAPE AND WEIGHT

Shape

Another contributory factor to the thermal efficiency of the building is its shape.

The smaller the external surface area of the building – its walls and roof – the less its heat loss. This is a directly proportional relationship.

Taking a typical UK, local authority, two storey house of 93 m² total floor area as a basis of comparison, it can be calculated that a square building with an overall height of 4.8 m has an external wall and flat roof surface area of 177.45 m². This figure can be considerably reduced if a hemispherical building of 126.27 m² floor area is substituted. The surface area then becomes approximately 139 m². There would, however, be practical difficulties in planning a Parker Morris dwelling within this shape, particularly bearing in mind the loss of floor area on the first floor, due to the reduction below 2 m of the ceiling height towards the perimeter. An increase in floor area of about 20% would probably be necessary to overcome this. The surface area of such a building would still be less than in the case of the square plan building (150.84 m²). Similarly, a circular building of 93.17 m² with vertical walls (a

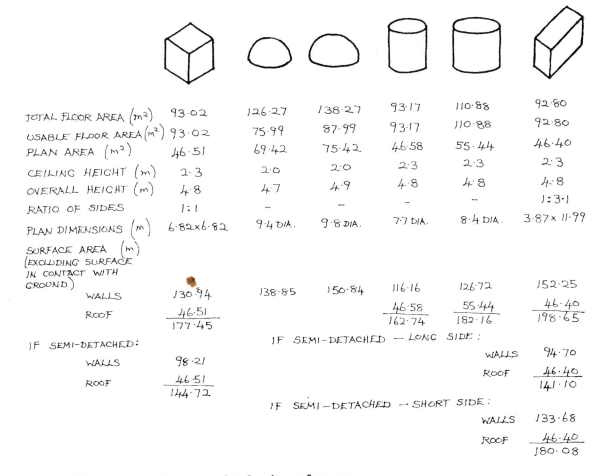

TOTAL FLOOR AREA (m²)	93·02	126·27	138·27	93·17	110·88	92·80
USABLE FLOOR AREA (m²)	93·02	75·99	87·99	93·17	110·88	92·80
PLAN AREA (m²)	46·51	69·42	75·42	46·58	55·44	46·40
CEILING HEIGHT (m)	2·3	2·0	2·0	2·3	2·3	2·3
OVERALL HEIGHT (m)	4·8	4·7	4·9	4·8	4·8	4·8
RATIO OF SIDES	1:1	–	–	–	–	1:3·1
PLAN DIMENSIONS (m)	6·82×6·82	9·4 DIA.	9·8 DIA.	7·7 DIA.	8·4 DIA.	3·87×11·99

SURFACE AREA (m)
(EXCLUDING SURFACE
IN CONTACT WITH
GROUND)

WALLS	130·94	138·85	150·84	116·16	126·72	152·25
ROOF	46·51			46·58	55·44	46·40
	177·45			162·74	182·16	198·65

IF SEMI-DETACHED:

WALLS	98·21
ROOF	46·51
	144·72

IF SEMI-DETACHED — LONG SIDE :

WALLS	94·70
ROOF	46·40
	141·10

IF SEMI-DETACHED — SHORT SIDE :

WALLS	133·68
ROOF	46·40
	180·08

Figure 3.3   The shape of a building related to its surface area

cylindrical shaped building) would have less surface area than a square building (162.74 m²), but would suffer from many of the disadvantages of a hemispherical building. It would, therefore, for practical purposes require an increase in floor area, when its surface area would exceed that of the square plan building.

Generally, it seems that any advantage of using a circular or quasi-circular plan form is outweighed by the resultant awkward shaped rooms which, because of their unaccustomed shapes, will probably include unusable space and, consequently, wasted floor area, since there is the problem of fitting rectangular furniture into non-rectangular rooms.

A rectangular dwelling of approximately the same area, the walls of which are in the ratio 1 : 3.1, would have a total external surface area of 198.65 m² – over 11% greater than that of the square plan building (Figure 3.3).

From this exercise it can be seen that the significance of differing side dimensions in simple, small scale buildings is relatively unimportant, and the effect of a larger wall surface area can be largely eliminated by the addition of a little extra insulation. However, it does not alter the fact that the more nearly a plan form approaches a square, the less is the heat loss from the building, all other facts being equal. This, therefore, is a point that cannot be ignored. The significance grows with the size of the building, or with a building having extremely disproportionate side wall ratios, or many set-backs in elevational line.

The effect of varying surface areas can be illustrated in similar sized domestic buildings. If a top floor flat is taken as representing the mean with regard to heat loss criteria, an intermediate floor gable flat would represent about 50% of that mean, a middle terrace house 94%, a semi-detached or end-terrace house about 120% and a detached house 151% (Figure 3.4).

The significance of the area of the roof in thermal assessment should not be ignored. Even in the UK, where the solar angle is 60°, an extensive flat roof can be an important solar energy collector. The southern slope of a pitched roof, which could be almost normal to the sun's rays, is an even more effective energy collector. As the building height increases (assuming the floor area remains constant) so the ratio of roof area to wall area decreases, reducing the potential solar gain – and also increasing the building's risk of an augmented air infiltration rate. However, this decreases the roof's winter-time heat loss potential, which is useful if the insulation characteristics are inadequate. This is just one example of the fact that thermal design is a fine balance of a series of conflicting pressures – summer and winter conditions, building size and shape, and thermal performance of the elements of the building shell.

In recent years a habit developed of designing office buildings in deep, open-plan form. These buildings usually had to be air-conditioned, because of the impossibility of adequately ventilating such plan forms by natural means. They also had to be artificially lit. There has been a great deal of discussion as to whether these buildings do, or do not, represent a wasteful use of energy.

There are many apparently good arguments for deep plan buildings. They have future adaptability without reconstruction, they help towards preserving the visual and social aspects of the urban scene, they provide a greater area of desirable working conditions for more people in less space, they have a smaller external surface area to floor area ratio. Clearly, the last two facts show positive advantages in producing a building of good thermal efficiency. The smaller the volume in which is created an artificial environment, the more economical it is. The less the surface area, the less the heat loss. Another thermal advantage is that, due to the elimination of opening lights, air infiltration (assuming well-designed wall components) can be almost eliminated.

The casual heat gain from artificial lighting, which is obviously essential in deep plan forms (usually about 40 W/m² for office buildings) together with heat gains

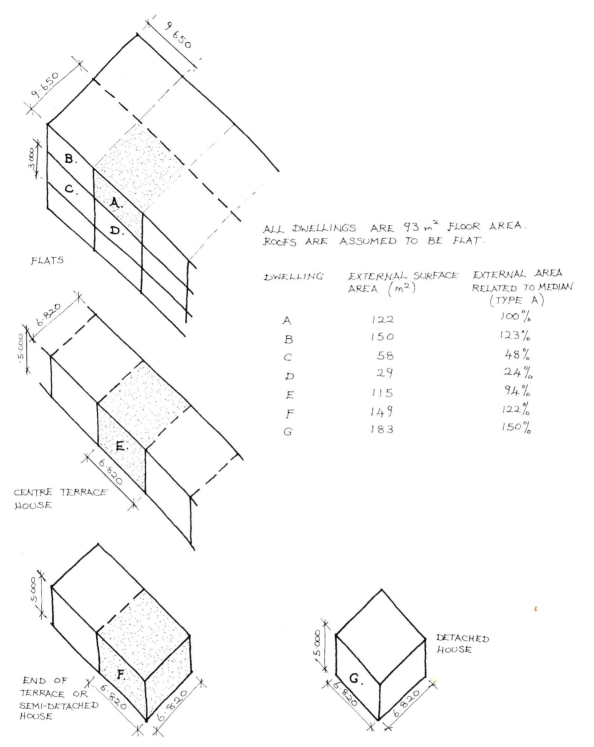

ALL DWELLINGS ARE 93 m² FLOOR AREA.
ROOFS ARE ASSUMED TO BE FLAT.

DWELLING	EXTERNAL SURFACE AREA (m²)	EXTERNAL AREA RELATED TO MEDIAN (TYPE A)
A	122	100%
B	150	123%
C	58	48%
D	29	24%
E	115	94%
F	149	122%
G	183	150%

Figure 3.4   External surface area of different dwelling types

from the occupants of the space and minor equipment (10 W/m²) is sufficient to remove the necessity for winter heating – the input adequately covering the fabric and ventilation losses. It does, nevertheless, demand mechanical ventilation at all times and considerable cooling load on the air-conditioning plant in summer. It has been suggested that if the building is correctly shaped to minimise external surface area, its elements have adequate thermal insulation properties and good thermal capacity, casual heat gain and acceptable solar radiation should be sufficient to render the internal environment practically self-sufficient for the majority of the year. Arguments based on this contention tend to ignore the total costs of running the environmental aids.

Table 3.3 Estimate of plant loads, hours of use and energy used with various building designs – with emphasis on conservation

Environment and building shape	Glazing window/wall %	Sun blinds	Glass	Plant details – W/m² floor				Annual use at full load-hours				Annual energy supply to building GJ/m² floor					Annual Primary Energy GJ/m² floor		
												Electricity				Heat			
				Refrig Power	Fans & Pumps	Lights	Boilers	Refrig	Fans & Pumps	Lights	Boilers	Refrig	Fans & Pumps	Lights	Total*	Boiler	for Elect	for Heat	Total*
Natural ventilation Natural lighting 13–15 m overall depth	50	Internal	Single	–	3	25	87.2	–	2000	1650	1000	–	0.02	0.15	0.17	0.48	0.64	0.52	1.16
	50	Internal	Double	–	3	25	66.2	–	2000	1650	1000	–	0.02	0.15	0.17	0.37	0.64	0.40	1.03
	50	External	Single	–	3	25	87.2	–	2000	1650	1000	–	0.02	0.15	0.17	0.48	0.64	0.52	1.16
	50	External	Double	–	3	25	66.2	–	2000	1650	1000	–	0.02	0.15	0.17	0.37	0.64	0.40	1.03
	30	Internal	Single	–	3	25	70.7	–	2000	1850	1000	–	0.02	0.17	0.19	0.39	0.71	0.42	1.13
	30	Internal	Double	–	3	25	58.0	–	2000	1850	1000	–	0.02	0.17	0.19	0.32	0.71	0.35	1.05
	30	External	Single	–	3	25	70.7	–	2000	1850	1000	–	0.02	0.17	0.19	0.39	0.71	0.42	1.13
	30	External	Double	–	3	25	58.0	–	2000	1850	1000	–	0.02	0.17	0.19	0.32	0.71	0.35	1.05
Air-conditioned Natural lighting 13–15 m overall depth	50	Internal	Single	21.9	15	25	58.9	1100	2700	1650	1000	0.09	0.15	0.15	0.38	0.33	1.43	0.35	1.78
	50	Internal	Double	21.9	15	25	37.8	1100	2700	1650	1000	0.09	0.15	0.15	0.38	0.21	1.43	0.23	1.65
	50	External	Single	15.8	15	25	58.9	1100	2700	1650	1000	0.06	0.15	0.15	0.36	0.33	1.34	0.35	1.69
	50	External	Double	15.8	15	25	37.8	1100	2700	1650	1000	0.06	0.15	0.15	0.36	0.21	1.34	0.23	1.56
	30	Internal	Single	17.1	15	25	42.4	1100	2700	1850	1000	0.07	0.15	0.17	0.38	0.24	1.42	0.25	1.67
	30	Internal	Double	17.1	15	25	29.7	1100	2700	1850	1000	0.07	0.15	0.17	0.38	0.17	1.42	0.18	1.60
	30	External	Single	13.4	15	25	42.4	1100	2700	1850	1000	0.05	0.15	0.17	0.37	0.24	1.37	0.25	1.62
	30	External	Double	13.4	15	25	29.7	1100	2700	1850	1000	0.05	0.15	0.17	0.37	0.17	1.37	0.18	1.54
Air-conditioned Artificial lighting 50 m square	50	Internal	Single	19.0	15	25	–	1800	2700	2200	–	0.12	0.15	0.20	0.47	–	1.74	–	1.74
	50	Internal	Double	19.0	15	25	–	1800	2700	2200	–	0.12	0.15	0.20	0.47	–	1.74	–	1.74
	50	External	Single	16.1	15	25	–	1800	2700	2200	–	0.10	0.15	0.20	0.45	–	1.67	–	1.67
	50	External	Double	16.1	15	25	–	1800	2700	2200	–	0.10	0.15	0.20	0.45	–	1.67	–	1.67
	30	Internal	Single	16.7	15	25	–	1800	2700	2300	–	0.11	0.15	0.21	0.46	–	1.72	–	1.72
	30	Internal	Double	16.7	15	25	–	1800	2700	2300	–	0.11	0.15	0.21	0.46	–	1.72	–	1.72
	30	External	Single	14.9	15	25	–	1800	2700	2300	–	0.10	0.15	0.21	0.45	–	1.68	–	1.68
	30	External	Double	14.9	15	25	–	1800	2700	2300	–	0.10	0.15	0.21	0.45	–	1.68	–	1.68

Assumptions: 1.0 air changes per hour
Coefficient of performance of cooling system is 3.5
2/3 of heat from lights in deep buildings is extracted with exhaust air

*Totals not always sum of parts due to rounding

Recent work at the BRE[1] has suggested that, in fact, deep plan office buildings can use as much as 60% more fuel in real terms than naturally lit and ventilated, traditionally planned offices. This assessment takes into account the very high electrical demands of the deep plan buildings. Because of the large energy loss in the conversion of the primary fuel into electricity, their demand on primary fuel is inflated. The conclusion seems to be, therefore, that the optimum building shape is one within which it is possible to make use of natural light and ventilation – a building of maximum depth 13 m–15 m. Comparative figures to support this view are given in Table 3.3. In this, narrow buildings naturally lit and ventilated are compared with similar shaped buildings naturally lit but air-conditioned, and deep plan buildings that are both artificially lit and air-conditioned.

Thermal weight

The secret of designing a thermally efficient envelope consists in matching the performance of the enclosing elements to the thermal demands which are going to be made on the building. A light, timber-framed building of low thermal capacity and adequate insulation will have a mercurial response to internal intermittent heat input, but almost as soon as the heat is turned off, in spite of its insulation, it will become cool; its comfort level is lost. This characteristic is common to all *lightweight* structures.

A *heavyweight* structure – one made of brick or concrete – is slow to respond to internal thermal input, but the heat it absorbs and stores can be used later on to even out the highs and lows of temperature, creating a more stable internal comfort temperature. This aspect of thermal performance will be discussed later when thermal admittance is examined; but in the vernacular examples of thermal design in this chapter the effect of thermal weight will be seen at work.

INTERNAL LAYOUT

Finally, the designer should turn his attention to the inside of his building and assess its internal layout for thermal efficiency.

When planning a building, the occupied spaces should be assembled around the heat source, just as in the vernacular English cottage rooms were planned around a central chimney breast.

Storage heaters should be placed on internal walls, the walls themselves (assuming they are of heavyweight construction) becoming extensions of the storage heater. The best heating services – and therefore the most effective in their use of fuel – are those in which the plant room has been planned so that the building can be served by direct, straight lengths of pipework or ducting. The plant room is not something to be stuck at the furthest end of the building – an architectural embarrassment to be hidden away from view. The services are not a magic plug-in feature appearing late in the design stage, but a fundamental aspect of design, thought about and discussed almost at the beginning of the design process.

The way in which internal spaces relate to openings in the building shell and the blocking of internal air infiltration routes are two things which should be considered (see Chapter 4).

The designer's objective is to plan a building of minimal volume, which effectively fulfils its required function. He should eliminate all waste space and should have an open mind concerning the revision of space standards. During the next few years it is likely that we shall see design minima being re-thought. The escalation of living standards and comfort standards is now over. We could be about to enter an era based on acceptable minima, and not desirable maxima. For instance, depending on the size of the room and its means of ventilation, the present mandatory minimum ceiling height for habitable rooms could well be reduced without endangering health. If the 2.3 m minimum height were reduced to 2 m, the ventilation heat losses could be reduced by 12½% and the heating costs similarly. This type of constructive thought is likely to become more familiar in the next few years.

VERNACULAR DESIGN FOR THERMAL COMFORT

To end this chapter three examples are given, from three different climatic zones, of traditional building techniques, applied by local craftsmen without very much consideration for what they were doing, but in the event producing buildings which were comfortable to live in without the aid of sophisticated mechanical plant. These are examples of thermally *passive* structures, and the present day designer could well spare time to discover why they were successful. He might then be able to apply some of the principles in his own designs.

The English country cottage (Figure 3.5)

    Climate: Often cold and damp.

    Problem: How to remain warm with a minimum of fuel.

    Siting: It is probable that, if the cottage were part of a village, the whole community would have sited itself not only where there was a functional reason for the growth of population, but also where advantage could be taken of any existing micro-climate – a sheltered valley, the leeward slope of a hill, or behind a natural wind-break. The layout of the buildings in relation to their site would be such that the main living rooms would be sheltered from the prevailing weather direction by outbuildings or barns, taking advantage of southerly aspects.

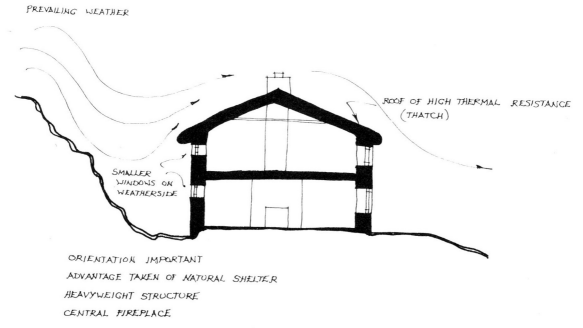

Figure 3.5 The traditional English cottage

    Shape: The cubic content of the dwelling would be kept to a minimum. Room heights were low. The use of the space within the pitched roof was an essential part of the living accommodation.

    Layout: Main living rooms were clustered around the heat source – the open fire and its associated chimney breast. Often this was centrally placed in order to take advantage of the four-sided radiation it afforded, in spite of the considerably greater difficulty of building a chimney in this position than on a gable wall. Main door openings tended to be either on sheltered elevations, or protected by porches.

    Construction: Walls were thick with good thermal capacity, although this was probably more to stop rain penetration than as the result of any thermal considerations. Windows were small and tended to be on the south side of the building. Roofs were pitched for weather protection and for the usable additional space they provided. Often roofs were covered by thatch – a particularly good insulating material.

The Middle East courtyard house (Figure 3.6)

> *Climate*: Hot and arid.
>
> *Problem*: How to remain cool in a hot and dry climate with clear skies.
>
> *Siting*: Where the building was positioned was usually relatively unimportant unless there were openings in the enclosure of the courtyard, in which case they were usually orientated towards the prevailing wind.
>
> *Shape and layout*: The main rooms were collected round the courtyard. This radiated heat to the cool night sky, and created a pool of cool air in the courtyard and adjoining ground floor rooms. During the day the inward-looking rooms were relatively unaffected by the sun beating on the outer walls, these having little or no glazing. A micro-climate was built-up inside the courtyard which the ground floor rooms shared – the effect amplified by the central pool of precious water which, in addition to its psychological benefit, produced a physical bonus by adding to the cooling through evaporation.
>
> *Construction*: Walls were thick with a high thermal capacity. The penetration of solar heat was therefore slow, and its loss during the night equally slow. In cold winter nights, this stored heat was welcome; in summer it was dissipated by natural ventilation when the ambient external air was cool. Walls and roof were painted white to reflect as much solar radiation as possible.

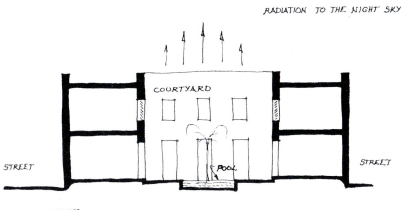

Figure 3.6   The traditional Middle East courtyard house

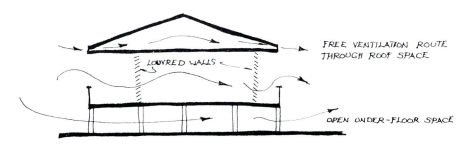

Figure 3.7   A typical tropical monsoon dwelling

Tropical monsoon dwelling (Figure 3.7)

Climate: Hot and humid.

Problem: How to remain cool in a hot and humid climate where there was no clear night sky to allow the radiation of stored heat and where high humidity inhibited evaporation cooling.

Siting: In this case the siting of the building was of the greatest importance. Every advantage had to be taken of the least air movement and the position of the building in relation to the prevailing wind was vital.

Shape and layout: As air movement and its direct cooling effect (as well as the cooling effect due to evaporation) was essential, buildings were one room wide, thereby allowing cross-ventilation. Large overhangs at the eaves provided shade from the direct rays of the sun and a shaded outdoor sitting area. Walls were constructed of louvres to allow maximum ventilation and ground floors were raised from the ground to allow free passage of air below.

Construction: High thermal capacity of the structure was relatively unimportant, since as there was no cyclic cooling there was no chance of disposing of the heat in store. Glazed areas were replaced by louvres to make the most of whatever air movement there was.

The three examples above serve to illustrate the unconscious use of some of the principles which have already been, or will later be, discussed. The miracle of our technological explosion has at times tended to obscure the fact that there were simple answers to problems in the past which still might have relevance today.

REFERENCE
1. Milbank, NO, *Energy consumption in 'other' buildings*, BRE, 1975.

4 The role of the building shell – exclusion and control of the external elements

The preservation of a stable internal environment in any building is one of the most costly operational items a building owner is going to have to finance. Therefore, it is to his advantage if the building shell can be designed in such a way as not to aggravate the already heavy cost of heating services by allowing unnecessary penetration by undesirable elements of the external environment. These could consist of air leakage and solar heat gain, though neither is wholly unwanted. Both have a contribution to make to the benefit of the internal environment, particularly in non-air-conditioned buildings, but their input must be controlled.

AIR INFILTRATION

Air infiltration is a source of heat loss and fuel wastage. It cannot be entirely eliminated, but it can be kept to a desirable minimum by thoughtful planning of the internal layout of the building, careful positioning of doors and windows, selection of windows of appropriate performance for the exposure of the site and skilful design of all joints in the external envelope.

Fortuitous air leakage – over normal levels expected and even required in naturally ventilated buildings – should not be allowed to occur. In air-conditioned buildings with non-opening windows, ideally there would be none. This is clearly unrealistic. A certain amount of air infiltration always occurs in practice, if only through doorways. Normally, some unintentional air leakage creeps through glazing gaskets or at curtain wall junctions when there is a large pressure differential between inside and outside environments. In the case of a naturally ventilated building with opening windows, the major source of air infiltration is at the joints between the opening lights of the windows and their frames.

A clear distinction needs to be made between natural ventilation, which is an intentional, controllable characteristic (and one that is very necessary to a comfortable environment) and infiltration which is an uncontrollable characteristic of the structure. It is possible in extreme cases for the infiltration alone to provide sufficient air changes for the adequate ventilation of the internal spaces. However, it is at the mercy of natural forces and cannot be relied upon. Under some climatic conditions the same leakage routes would be totally insufficient. The same applies to the effectiveness of openable windows, which under certain circumstances can prove inadequate.

For air infiltration to occur there needs to be pressure differential between the inside and the outside of the building. This can be the result of:

1. wind on the external face of the building, or
2. the buoyancy of air, resulting in an upward drift of air between gaps at different heights – a characteristic known as *stack pressure*.

Wind pressure

Any square-sided building presents an obstacle to the wind, which produces complex changes to its flow pattern involving local fluctuations of velocity and pressure (Figure 4.1). Wind blowing squarely on one elevation will produce positive pressure on that face and negative pressure (or suction) on all remaining faces. If the wind is blowing on a corner of the building, the two windward faces will be subjected to positive pressure, the two leeward faces to negative pressure.

Pressure can be calculated from the following formula which appears in BS CP 3: Chapter 11: 1970:

$$P = 0.623 \ v^2$$

in which P = pressure (N/m²) and
v = wind velocity, being derived as described in Chapter 3, page 16.

In spite of local pressure variations, and although on windward facades pressure can become quite severe, it is usually sufficiently accurate, when considering the total building envelope performance, to consider only the *mean surface pressure*.

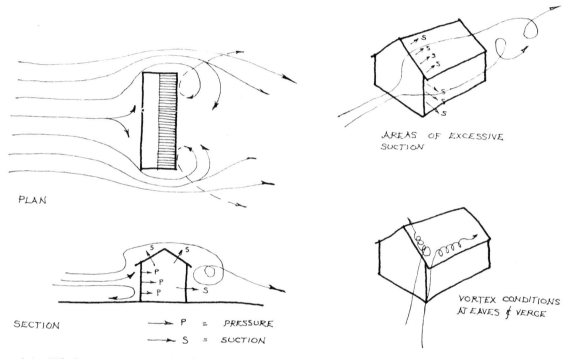

Figure 4.1   Wind patterns around buildings

Most buildings are permeable to some extent, either due to windows, ventilators, or gaps (nominally sealed) in the structure. These apertures cause air paths to be built up between opposite or adjacent faces of the building experiencing different air pressures. The ventilation rate is then approximately in proportion to the area of the gaps and the square root of the pressure difference.

Gap area is assessed by the product of the average width of the gap and the length of the gap. Areas of gaps in parallel (i.e. on the same face or elevation) can be totalled to arrive at the gross gap area. Gaps in series (i.e. in opposing facades or in all the partitions between the facades) produce a greater resistance to the passage of air. Their total effect is a gap area of rather less than the smallest gap on the air path. The whole problem is complex and one that need not concern the building designer in detail. However, fundamental design recommendations emerge which will help the designer to anticipate the performance of his building in respect of air infiltration.

Stack pressure

Ventilation, or external air infiltration, can be produced by air buoyancy – the natural characteristic of air to rise if there is a temperature difference involved. Stack effect particularly operates in heated buildings where gaps in the structure occur at different levels. Flow is induced from the lower to the upper gaps in a similar fashion to the natural ventilation flow up an unused domestic flue. The pressure developed is directly proportional to the vertical distance between the gaps and the mean difference between indoor and outdoor temperature. No wind is necessary for stack pressure to develop; it is merely a physical characteristic of the building shape and vertical pathways within the building.

It is not usual for either wind or stack pressure to be present on its own. Only in very abnormal conditions (mild weather with strong winds, or very cold weather with no wind) does one pressure operate exclusively (in the first case, wind pressure; in the second, stack pressure). Otherwise, the two effects operate together, although usually one tends to dominate. In summer, when there is the slightest chance of the smallest wind, stack pressure can generally be ignored. In winter the effect of flues, stairwells and other vertical paths makes stack pressure significant, and adequate enclosure of these elements is necessary in order to control unanticipated negative pressure inducing increased air infiltration, which in turn would have marked effect on the required heating output.

DESIGN RECOMMENDATIONS

Precise calculation of the air infiltration rates of a building can be made by applying the principles set out in BRE Digests 119 and 210. These are related and refer to BS CP 3: Chapter V: Part 2: 1972. This is, however, a relatively complex process and one that would normally be beyond the average designer's day-to-day experience. The services engineer should obviously be involved to advise on this matter and, at an early design stage, on other aspects of the shell's performance. The designer should, though, be aware of the principles involved – and the effect of applying these principles. For this reason a set of design recommendations is given here, together with other simple methods of ensuring the correct selection of adequate windows and of generally evaluating air infiltration risks.

Fundamental design recommendations are:

1. The building envelope should be made as airtight as possible to avoid fortuitous air infiltration. An assessment of the anticipated wind pressure can be made using local mean wind speed information, or the BS CP 3: Chapter 11 method. This is then applied to the formula quoted earlier. From this a comparative judgement of the exposure of the site can be made, and the designer will know whether the joints of his building envelope are to be subjected to a greater than average, average, or minimal level of risk. His detailing should then be adjusted accordingly to take account of this. It is worth remembering that wherever air leakage occurs in a building shell, there is a real danger (unless the geometry of the joint and gravity prohibit it) of water following the same path[1]. Even in open drained joints it is necessary for there to be an airtight seal at or near the inside face of the joint.
2. Care must be taken in the design of doors and windows, since clearly the point of major air infiltration will be at such openings in the envelope. (These will be discussed later in greater detail). The number and position of doors will be dictated by the requirements of access to and escape from the building, in addition to its internal layout and circulation patterns. However, the careful use of draught lobbies and the avoidance of through-foyers planned with access doors on opposite elevations (particularly when associated with lift or stairwells) will help to minimise the problem. The number of openable windows in naturally ventilated buildings will be dictated by ventilation requirements of the building (usually 5–6 litres of air per person for sedentary areas where smoking is prohibited, 12 litres where smoking or greater activity levels are anticipated, even higher inputs where there is the presence of dust or fumes). Openable windows, however, should be kept to a minimum consistent with adequate

ventilation, bearing in mind that a too conservative approach can lead to periods of discomfort due to poor internal distribution of fresh air, or dependence on natural forces to achieve the required ventilation rates; such a period would occur during hot, still summer weather.

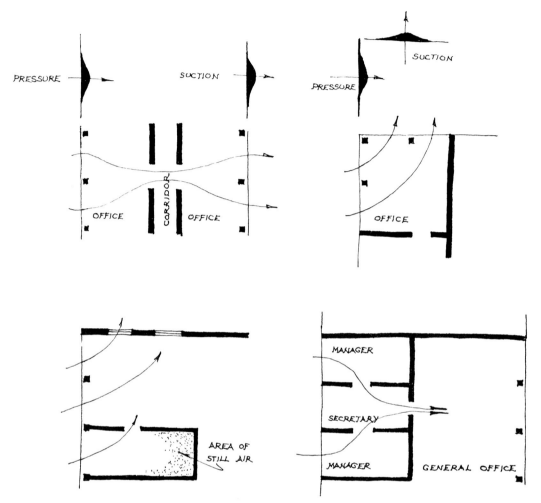

Figure 4.2   Internal air paths inside a building

3. Take cross ventilation into account. Whilst this is a necessity in naturally ventilated buildings in hot climates – and even on some hot, still days in the UK – rooms with windows on two opposite elevations are subject to greater air infiltration due to the absence of partitions in the line of the air paths. The same comment applies to open-plan offices with an absence of full height partitioning. Figure 4.2 shows typical layouts and their effect on air infiltration rates. Further amplification of this point will be given later in this section.
4. Consider other minor openings in the external shell which contribute to air infiltration – air bricks, ventilation grills, etc. – and do not overlook their effect when assessing air infiltration. It is likely that in the future there will be a more lavish provision of permanent ventilation, particularly in domestic buildings, with the object of assisting in the reduction of condensation, but the by-product will be increased heat loss and fuel consumption. This will be discussed in the second part of this book.

Choice of window   The chief flow of unanticipated air infiltration of the building shell is through the gaps in windows. For this reason their selection is a matter of considerable importance, and money spent initially on a suitable quality of window, whose performance matches the exposure of the proposed location, is money well spent.

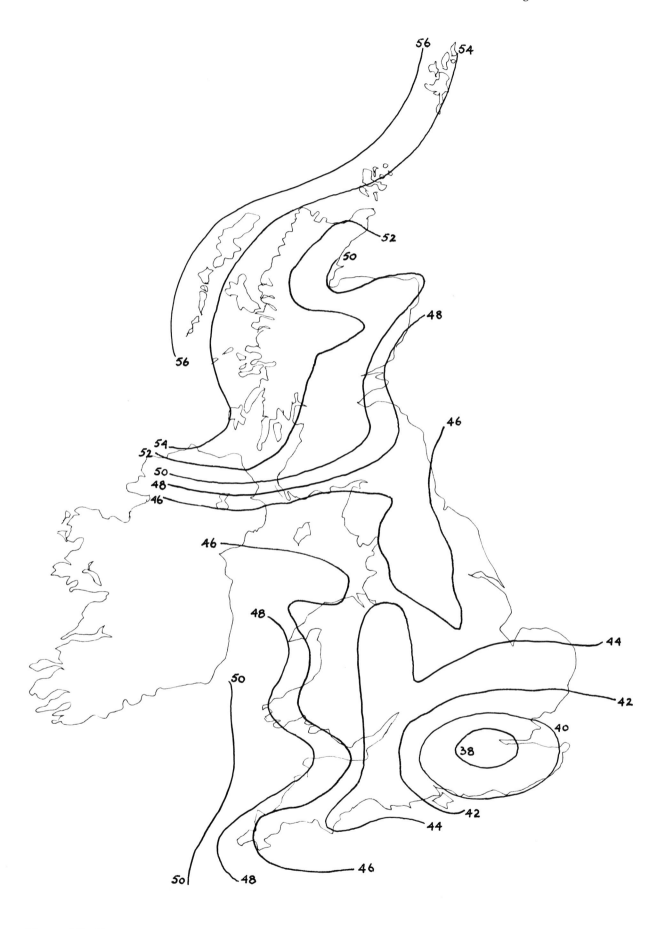

Figure 4.3   Exposure zones

The documentation of performance standards for windows with regard to air (and, for that matter, water) leakage has been confused. On the one hand there was a BS test method, which was unrelated to performance standards; on the other hand, Agrément Board tests which were so related. The situation was further confused by another set of slightly different standards laid down by the Interdepartmental Construction Development Group of the DOE, DES, DHSS and SDD[2]. At the time of writing, the BSI are working on a new Standard which will supersede the previous Draft for Development DD4:1971. This Standard should resolve the confusion that exists at present. Some levels of acceptance are likely to be modified as a result of comparison between laboratory test results and actual site performance, and in order to achieve harmonisation with European standards no doubt the Pascal will be used as a unit of pressure (1 Pa = 1 N/m^2 = 0.102 mm H$_2$0).

Window infiltration coefficient for 1 N/m^2 (litres/ms)	Window type
0.05	Horizontally or vertically pivoted – weather stripped.
0.125	Horizontally or vertically sliding – weather stripped.
0.25	Horizontally or vertically pivoted – non-weather stripped. Horizontally or vertically sliding – non-weather stripped.

Table 4.1   Air infiltration categories through windows

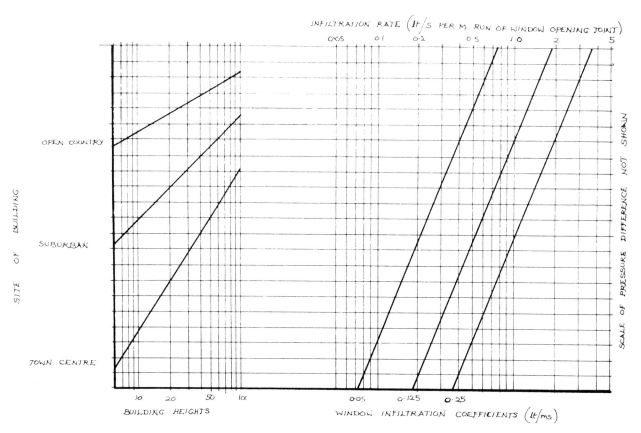

Figure 4.4   Air infiltration through windows

Windows will probably be subdivided to those intended for *normal exposure* and those intended for *severe exposure*, with tables and charts defining both the exposure zones and also the heights of buildings in different locations considered as experiencing one or other of the two grades of exposure. Figure 4.3 gives an indication of the differing conditions experienced in the UK, the isopleths representing basic wind speeds (maximum 3 second gust speed likely to be exceeded only once in 50 years at 10 m above ground in open, level countryside) at 2 m/s intervals.

In the new Standard it is likely that tables of design wind pressure will be given, based on building shape and ground roughness characteristics, and that window manufacturers will be required to state that the weather tightness of their product will meet the requirements for a design wind pressure of, say up to 2000 Pa or of a specified pressure above 2000 Pa. The latter would clearly be high performance windows.

GRAPHICAL METHOD OF ESTABLISHING ANTICIPATED AIR INFILTRATION

The CIBS Guide gives a ready method of calculating the total air infiltration likely to be experienced by buildings of various heights in three types of location – town centre, suburban and open country. Windows are divided into three categories (Table 4.1) and these and the three types of site are plotted over a grid of pressure difference based on meteorological wind speeds to produce an infiltration chart (Figure 4.4).

This chart is used in the following manner:

First the pressure difference across the building is established. Fix the building height (say 50 m) on the bottom scale of the left hand side of the chart and project a line upwards until it intersects the appropriate site line (say suburban). From this point draw a horizontal line to the right until it intersects the required window category (say 0.05 litres/ms). Then project a line upwards to the top scale and read off the infiltration rate – 0.35 litres/s per m run of opening joint.

This rate has then to be modified to take into account the amount of internal obstruction to the passage of air from one side of the building to the other – the amount of partitioning – and the amount and quality of the windows on the elevations. This is done by simply using the following equation and applying in it a correction factor found in Table 4.2:

$$Q_b = Q \times f$$

in which Q_b = basic infiltration rate (litres/ms)
Q = infiltration rate from the infiltration chart (litres/s)
f = correction factor from Table 4.2.

Q_b is then multiplied by the opening joint length for any particular room or internal space and Q_r (the room infiltration rate) is established:

$$Q_r = Q_b \times L_r$$

in which L_r = opening joint length per unit area (m/m²).

In multi-storey buildings an adjustment may have to be made to allow for the stack effect. In winter the lowest floors can experience above average air infiltration, the topmost floors below average. In summertime the situation can be reversed (see Table 4.3).

Finally the total air infiltration (Q_{tot}) can be calculated from the following formula:

$$Q_{tot} = Q_b \times L_r \times A_{rep}$$

in which A_{rep} = a representative area.

35

Window type	Internal structure	Correction factor (f)
All types	Open plan (no full partitions)	1.0
Short length of well-fitting window opening joint (approx. 20% of facade openable)	Single corridor with many side doors: liberal internal partitions and few interconnecting doors	1.0
Long length of well-fitted window or short length of poor fitting window (approx. 20–40% of facade openable)	Single corridor	1.0
	Liberal partitions	0.8
Long length of poor-fitting window joint (approx. 40–50% of facade openable)	Single corridor	0.8
	Liberal partitions	0.65
Very long length of poor-fitting window joint (in excess of 50% of facade openable)	Single corridor	0.65
	Liberal partitions	0.4

Table 4.2 Correction factors for use in assessing window infiltration

Condition	No. of storeys	% increase in infiltration above average	Level of maximum ventilation
Wind acting alone	5	3 ⎫	Topmost floor
	10	6 ⎬	
	20	8 ⎭	
9 m/s wind plus stack effect (20°C heating season)	5	3 ⎫	Lowest floor
	10	10 ⎬	
	20	20 ⎭	

Example: In winter the ground floor of a 20 storey building experiences a 20% increase on average infiltration rates. This excess decreases linearly to zero at mid-height.

Table 4.3 Percentage deviation from average infiltration rate experienced in multi-storey buildings

It should be remembered that the total air infiltration, being due to differential pressure, largely consists of air entering a building on one side and passing out on the other. Therefore, cold air ingress takes place only over an area equivalent to one elevation in a building with windows only on two opposite sides, or equivalent to a diagonal plane across the building when all elevations are glazed (Figure 4.5).

EMPIRICAL VALUES FOR AIR INFILTRATION

The CIBS Guide also lays down empirical values for air infiltration for use before the building design has sufficiently progressed for calculations to be undertaken. These values are included in Table 2.2 and have been devised for various types of building of typical construction in a *normal* exposure. It has been assumed that the average ratio of openable area (doors and windows) to wall area will be 25%. If this ratio is exceeded on one external wall, the infiltration rates must be increased by 25%; if on two or more external walls, by 50%. On severely exposed sites, rates should be increased by 5%; but on sheltered sites reduced by 33%.

For the purpose of this empirical method, exposures are defined as being:
Sheltered: Up to the 3rd floor of buildings in city centres.
Normal: Most suburban and country premises; fourth to eighth floors of buildings in city centres.
Severely exposed: Buildings on the coast, or on a hill site; floors above the fifth of buildings in suburban or country districts; floors above the ninth in city centres.

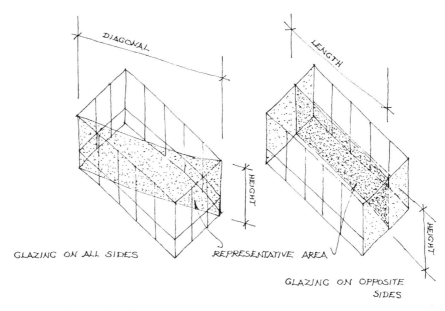

Figure 4.5 Representative area

SOLAR HEAT GAIN

In the tropics the sun, for very good reasons, is treated with respect; much more respect than it is given in temperate parts of the world, such as the sun-starved UK, where sunshine is either ignored completely, or is treated as a totally beneficent god by its hordes of worshippers. As a potential threat to the internal environment, it is largely dismissed in temperate zones. The result can be some very uncomfortable peaks of temperature behind over-large windows – and not only on hot summer days.

In frosty, clear, winter weather comfortable heating levels can be ruined by radiation from low altitude sunshine, producing unexpected peaks of temperature that can result in substantial discomfort for the occupants of the rooms on the sunny side of the building. Ironically, it can cause equal, but contrasted, discomfort for those on the shaded side, due to the lack of flexibility in a heating system which is unable to balance the disproportionate demands for heat of the two sides of the building. In air-conditioned buildings the solar heat gain through badly orientated large windows can place a wholly unnecessary load on the plant, and hence can cause fuel wastage. In winter, such uncalled-for solar heat gain causes windows to be flung open to dissipate the unwanted heat from the radiators – another criminal waste of fuel.

In short, the sun cannot be ignored, even in the UK. In controlled moderation, it is a positive benefit; uncontrolled, with a badly orientated building, it creates an untenable situation.

The upper limits of the earth's atmosphere receive a mean value of solar radiation of about 1395 W/m² and this is taken as the *solar constant*. The precise amount of solar radiation is varied by sun-spot activity and sun to earth distance. About 30% of this radiation is reflected back into outer space without change of wave length. The rest is absorbed by the atmosphere and the earth's surface, causing a rise in temperature, generating currents, waves and winds, forming our climate and causing the hydrological cycle. A small part enters the biological system through photo-synthesis in plants and other 'producer' organisms.

Atmospheric absorption reduces the intensity of radiation, depending partly on the length of travel through the atmosphere (the lower the sun's altitude, the weaker the radiation) and partly on the state of the air mass (cloudiness or floating dust). Nevertheless, at a zenith position, the intensity of radiation on a horizontal plane at sea level may approach 1 kW/m². The annual total amount of radiation received clearly depends on geographical latitude and local climatic factors.

In UK buildings, solar heat gain is mostly a characteristic of glazed areas of the building shell. It is rarely that the opaque parts of the normally insulated and relatively high thermal capacity shell will allow excessive transmission of external heat to the interior spaces. This is not so in the tropics. Here, due to the vernacular habit of shading windows and restricting their size, heat gain through glazed areas is often less important than through the rest of the building envelope. The principles of solar heat gain through opaque walls and roof, therefore, will not be ignored in this chapter, although they have little significance in UK design.

Through glazed areas   The problem of solar heat gain through glazing has been exacerbated in the last thirty years by the so-called modern international movement in architecture, which

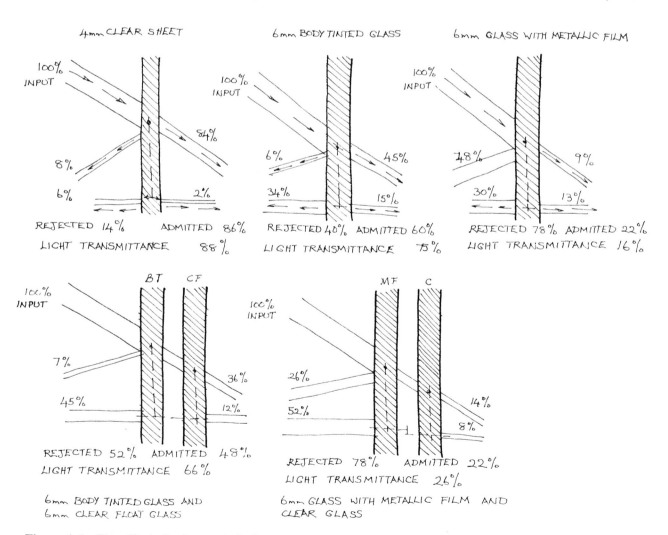

Figure 4.6 The effect of solar control glass

made a cult of the 'skin and skeleton' building to the extent that all walls were in danger of becoming made entirely of glass – or at least lightweight, low thermal capacity infill between mullions and transoms of steel or aluminium. This has had some very undesirable repercussions on the internal environment, associated not only with solar heat gain, but also with heat loss. In addition, high levels of natural illumination can lead to unpleasant glare internally.

Some designers, whilst clinging to the fashionable aesthetic, have resorted to double glazing and the use of heat reflecting or absorbing glass in an effort to ameliorate the thermal disadvantages.

The use of solar control glasses certainly has the effect of diminishing solar heat gain. Figure 4.6 indicates diagrammatically how some of these glasses work and compares their performance with that of clear sheet glass. Of the radiation falling on glass, even on 4 mm clear sheet glass, some is transmitted to the interior, some reflected to the outside and some absorbed, to be later either re-radiated or convected outwards or inwards. The proportions applicable to different glass types and glazing systems are illustrated. Of the heat absorbed by the glass, the proportion which is released indoors is indicated by the *partition factor* (P). Total heat admitted (F) is therefore:

$$F = T + P \times A$$

in which T = the directly transmitted component and
A = the solar energy absorbed by the glass.

The values quoted in Figure 4.6 are for the whole of the solar radiation spectrum – ultra-violet, visible and infra-red. About half the energy is in the visible part of the spectrum; hence, if the transmittance of the solar control glass is less than 0.50, the light transmittance is less than that of clear glass. No matter how the glass works, whether it reflects part of the solar energy from a thin metallic layer within or on the glazing system, or from metal ions injected into the glass surface, or absorbs the heat and achieves its effect by re-radiating the heat outwards, all methods result in a loss of light transmittance. In many ways these sophisticated glasses merely allow the designer to have bigger windows which achieve the same light and heat transmittances as smaller windows. Clearly, from the illustration, the double glazed systems can be seen to give a better performance than similar glass used as single glazing. The thermal transmittance from inside to outside is also significantly reduced. But special glass and double glazing are both very expensive, and not entirely efficient, means of correcting what may not necessarily have needed correction in the first place.

It has been reported from the USA that the arrival of a new building, entirely clad in solar control glass, has been known to reflect so much heat in the direction of its neighbours that it completely upset their internal environments, placing massive new demands on their air-conditioning plants which could not be satisfied. Certainly this provides an interesting, new, legal talking point!

The plain fact is, that if a thermally efficient building is to be designed solar heat gain has to be considered from the outset. Orientation plays a very large part in the avoidance of future trouble. Control of solar input is not just a matter of bolting on a few sunshades (albeit these can prove very effective, as we shall see later) or, for that matter, glazing with some sophisticated glass. It is as basic as how to position the building on the site, and where and what size the windows are to be.

THE PRINCIPLES OF SOLAR HEAT GAIN

In temperate climates, levels of solar heat gain likely to be experienced by a building need to be assessed at the time of initial design to establish whether the site orientation, the window sizes and their positions have a beneficial effect on the internal environment in winter as well as in summer, or could be improved by re-orientation, re-fenestration or by employing shading devices. The majority of buildings will acquire the whole of their summer heat input from casual gain from occupants, lighting or equipment, and from solar radiation. Therefore the time of day when radiation is to be experienced needs to be matched with anticipated occupancy patterns of the building. Also, the means of ventilation and its control need to be assessed in relation to positions in the building where peak radiation may be experienced. In air-conditioned buildings the summer peak temperatures need to be calculated, including solar radiation, in order to establish the maximum cooling load required from the plant.

To the above considerations, in hot climates, needs to be added the effect of solar heat gain through the opaque parts of the building shell. Even in more temperate climates this aspect cannot be wholly ignored, particularly in the case of low thermal capacity structures with poor insulation.

The level of solar heat gain experienced by a building is dependent on the following factors:

1. The angle of incidence of the sun's rays on the building. This is controlled by the sun's altitude (height above horizon), the angle of solar azimuth (the angle of the sun measured on a horizontal plane clockwise from the north) and the angle between the solar azimuth and a perpendicular to the building plane (angle of solar/wall azimuth). (See Figure 4.7).
2. The proportion of glass to opaque parts of the building envelope. As we have seen, glass is almost wholly transparent to solar radiation and therefore allows practically its full intensity to pass inside instantaneously. This effect can be modified by sun-shading devices.
3. The proportion of radiation absorbed and transmitted through the opaque structure, or through blinds to the glazed areas. Heavyweight walls and roofs of a high thermal capacity absorb some radiation. The heat will then pass slowly to the interior. A long-standing rule-of-thumb was that every 25 mm thickness of brickwork or masonry would produce one hour's time lag in heat transmission. Hence a wall 300 mm thick would impede heat flow for twelve hours – long enough for the radiation peak to have long since passed and the external temperature cooled – at which time the heat would commence to make its way back out again.

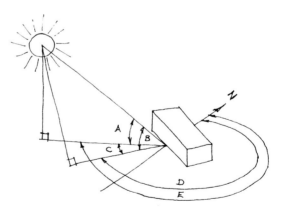

A = SOLAR ALTITUDE
B = ANGLE OF INCIDENCE
C = WALL / SOLAR AZIMUTH
D = WALL AZIMUTH ANGLE
E = SOLAR AZIMUTH ANGLE

Figure 4.7 Solar angles

The angle of incidence   In UK latitudes the intensity of solar radiation grows rapidly during March and dies at a similar rate after September. This is due to the differing solar altitudes and the varying thickness of atmosphere through which the radiation has to pass. In spite of this, the higher the solar altitude, the more oblique the angle of incidence of radiation becomes on south-facing wall surfaces and, therefore, the less, proportionately, the solar heat gain. East and west elevations, however, experience increasing radiation from March to September, because the solar altitude is higher at the time of day when the angle of solar/wall azimuth is smallest. The angle of solar incidence on southerly elevations during winter months is lower and so they experience proportionately greater heat gain at these times – a fact that, if appreciated, can be used to advantage to reduce winter fuel consumption. Generally, throughout the year, south-facing elevations experience more consistent solar radiation than any other elevations except north elevations, that receive no direct radiation whatsoever.

Figure 4.8 shows a month-by-month graph of average daily heat transmittance through vertical glazing. It illustrates clearly the effect of the obliqueness of the angle of incidence on solar radiation. From this it can be seen that a building with windows predominantly on the north and south elevations, rather than the east and west, will be less sensitive to solar gain and more consistent in the amount of solar input it receives throughout the year.

The effect of the sun position in the sky and the relationship to the siting of a building is demonstrated in the sun chart.

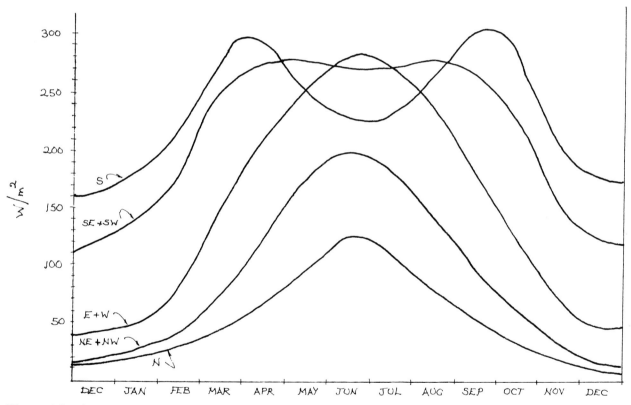

Figure 4.8 **Average daily rate of heat transmission through vertical single glazing in clear weather**

The sun chart

This is a useful graphical device for assessing the relationship of the proposed building and the path of the sun throughout the year. Figure 4.9 shows a typical sun chart drawn for latitude 51.7° North and therefore applicable to the UK.

The method of drawing a sun chart is as follows: first, a circle of convenient radius is drawn; then, starting from the top, this circle is divided into 10° segments representing lines of azimuth. In one quarter of the circle, from the points where the lines of azimuth cut the circumference, perpendiculars are dropped to meet the horizontal diameter. Concentric circles are then drawn, using the centre of the original circle and with circumferences passing through the meetings of the perpendiculars and the horizontal diameter. These circles represent solar altitudes from 0° to 90°. By reference to tables of solar altitude and azimuth, included in the CIBS Guide, for the appropriate latitude and time of year points can be plotted on the chart for the position of the sun at different times of the day (sun time – or GMT in UK – is used). Connecting these points with a smooth line produces an illustration of the sun's path during a particular month. The plan of the proposed building can then be added at the centre of the chart, correctly aligned. The effect of the sun's path during the course of the year on the radiation falling on the various surfaces of the building is clearly illustrated.

The sun chart shows at the same time which faces of the building are receiving direct radiation at any particular hour and also the wall/solar azimuth. By drawing a line from the sun position through the wall being considered to the centre of the circle, the wall/solar azimuth is the angle between this line and a line drawn normal to

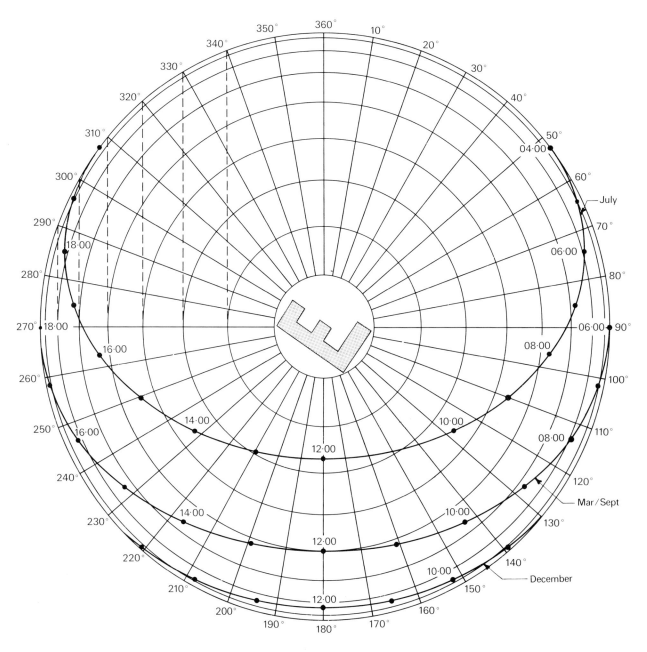

Figure 4.9   Sun chart for latitude 51.7°N

the wall. Any wall which is not cut by a line drawn from the sun position to the centre of the circle is in shade.

The obliqueness of direct solar radiation, as indicated above, significantly affects the amount of radiant heat experienced by various planes of the building. Also the intensity of direct solar radiation on cloudless days is directly related to the thickness of the atmosphere through which it has to travel. The lower the solar altitude, the greater the thickness of atmosphere, and hence the greater the dissipation. These facts are illustrated in Table 4.4.

	Wall/sun Azimuth	Sun altitudes									
		5	10	20	30	40	50	60	70	80	
VERTICAL WALL	0	210	382	584	642	624	553	447	312	160	Direct solar radiation (W/m²)
	10	207	376	575	632	615	545	440	307	158	
	20	197	360	550	603	586	520	420	293	150	
	30	182	330	506	556	540	480	387	270	140	
	40	160	292	447	492	478	424	342	240	123	
	50	135	246	375	413	400	355	287	200	103	
	60	105	190	292	210	312	277	224	156	80	
	70	72	130	200	220	213	190	153	107	55	
	80	36	66	100	110	108	96	78	54	28	
		15	22	31	37	42	46	50	53	56	Diffuse radiation (W/m²)
HORIZONTAL ROOF		18	67	212	370	523	660	773	857	907	Direct radiation (W/m²)
		30	43	62	75	84	92	100	107	113	Diffuse radiation (W/m²)
		24	55	137	222	303	376	436	482	510	Ground reflected (walls only)

Table 4.4   Intensity of solar radiation on clear days on sites 0 to 300 m above sea level

The direct radiation on a wall with a wall/solar azimuth of 0° (i.e. the sunlight is falling normal to the wall) will increase from 210 to 642 W/m² – a factor of 3 – between solar altitude 5° and 30° (see Table 4.4). The fall-off from 30° to 80° is due to the more oblique incidence of the radiation.

The impact of obliqueness of incidence is further illustrated in the direct radiation experienced by horizontal roofs. In this case, the direct radiation increases consistently with the solar altitude. The range is substantially greater. The effect of obliqueness here is accentuated by the filtering of low altitude radiation.

It should be remembered that amounts of radiation are affected too by the altitude of the site – Table 4.4 was drawn up for sites between 0 and 300 m above sea level – and the amount of water vapour, dust and ozone in the atmosphere, which tend to scatter and absorb radiation.

Heat from the sun reaches the earth by diffuse radiation, as well as the more obvious direct path. When radiation from the sun reaches the earth's atmosphere some of it is scattered, either back into space or down to the earth. It is this scattered radiation that produces heat on cloudy days. Diffuse radiation is also subject to the filtering effect of length of travel in the atmosphere (see Table 4.4) and is therefore proportional to the sun's altitude (i.e. it is unaffected by the wall/solar azimuth). On partly cloudy days the total of the sun's radiation is dramatically reduced; but it is the direct component which is cut, the sky diffuse intensity then being higher than on cloudless days.

A certain amount of heat, particularly in hot climates, is reflected from the ground. These intensities are included in Table 4.4 and are, as would be expected, directly proportional to the sun's altitude.

It has been found that certain correction factors need to be applied to direct radiation and ground reflectance figures, depending on climatic zone and clarity of sky. These correction factors are given in Table 4.5 and are to be used in calculating total solar intensity.

Situations	Direct radiation factor k_c		Ground reflectance factor k_r
	rural	urban	
Temperate zones	0.95	0.90	0.20
Tropical, arid	1.20	1.10	0.50
Tropical, humid	1.00	0.95	0.20

Table 4.5 Correction factors for direct radiation and ground reflectance

Inclined surfaces, such as the planes of pitched roofs, receive direct radiation at a rate dependent on their angle of inclination relative to the sun's rays. Obviously the more nearly normal to the radiation the pitch of the surface, the greater the radiation. A method of readily establishing a correction factor for various slopes of roof is to build up a figure based on combining two percentages, one from the radiation received on vertical walls, the other on horizontal roofs. Clearly the steeper the slope, the more significant becomes the solar radiation on the vertical surface and the less on the horizontal. Table 4.6 indicates the percentages.

Inclination of roof to horizontal (degrees)	Position of the sun with respect to roof surfaces					
	on same side of ridge			on opposite side of ridge		
	% intensity on horizontal roof		% intensity on vertical wall	% intensity on horizontal roof		% intensity on vertical wall
5	99.6	plus	8.7	99.6	minus	8.7
10	98.5	plus	17.4	98.6	minus	17.4
15	96.6	plus	25.9	96.6	minus	25.9
20	94.0	plus	34.2	94.0	minus	34.2
25	90.6	plus	42.3	90.6	minus	42.3
30	86.6	plus	50.0	86.6	minus	50.0
40	76.6	plus	64.3	76.6	minus	64.3
50	64.3	plus	76.6	64.3	minus	76.6
60	50.0	plus	86.6	50.0	minus	86.6
70	34.2	plus	94.0	34.2	minus	94.0

Table 4.6 Intensity of direct solar radiation on inclined roof surfaces with a clear sky

CALCULATING THE INTENSITY OF SOLAR RADIATION

The total intensity of solar radiation received on vertical surfaces may be calculated by the following formula:

$$I_t = k_a [(I_v k_c) + I_d + 0.5(I_h k_r)]$$

in which I_t = total intensity of radiation (W/m²)

I_v = intensity of direct radiation on vertical surfaces (W/m²)

I_d = intensity of diffuse radiation (W/m²)

I_h = intensity of total radiation on horizontal surfaces (W/m²)

k_a = correction factor for height (established from graph on Figure 4.10)
k_c = correction factor for sky clarity (Table 4.5)
k_r = correction factor for ground reflectance (Table 4.5)

A list of solar intensities on vertical and horizontal surfaces throughout the year and for latitude 51.7°N (UK) is given in Table 4.7.

Glazed and opaque areas of wall

Earlier in this chapter the reaction of various forms of glazing to solar radiation was examined and it was seen that, no matter what the method of solar control, unshaded, transparent parts of the wall, if they received any direct radiation, would let in the heat. In other words, if they were facing the southern segment of the sky in the northern hemisphere or the northern segment of the sky in the southern hemisphere, they would let in the solar radiation more or less in proportion to the amount of light they let in. We also observed that the obliqueness of the solar radiation on south-facing windows in northern latitudes tended to decrease their summertime reception of heat, while east and west aspects in both hemispheres are more vulnerable to radiation input.

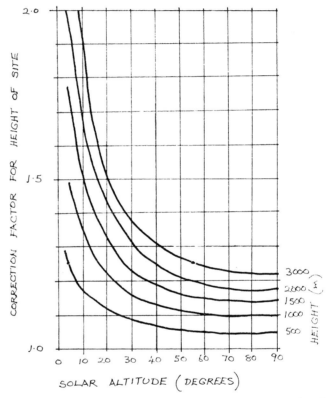

Figure 4.10 Variation in solar radiation with height above sea level

Recent work at BRE[3] analysed the window area necessary in naturally lit and ventilated office buildings. This indicated that buildings with a depth of 13 m –14 m require a percentage of window to wall area of between 30% and 50% to produce acceptable illumination levels. These ratios correspond to glass/floor area ratios in the range 0.13 to 0.22. The impact of this amount of glazing on internal comfort was then examined. Two standards of comfort were set – 'satisfactory', which assumed the wearing of lightweight suits, and 'minimum acceptable', in which the wearing of only shirt and trousers was comfortable.

Table 4.8 displays the results. These show that rooms with unprotected glass, when exposed to the sun, are not acceptable by either comfort norm – in fact, only the use of external shading would produce 'satisfactory' conditions. On the other hand, 'minimum acceptable' conditions can be achieved by the use of either internal or external shading, so long as it does not obstruct the ventilation air flow.

Date	Orien-tation	Daily mean	Sun Time 04.00	05.00	06.00	07.00	08.00	09.00	10.00	11.00	12.00	13.00	14.00	15.00	16.00	17.00	18.00	19.00	20.00
June 21	N	90	45	180	190	115	95	110	125	135	135	135	125	110	95	115	190	180	45
	NE	140	75	360	510	530	460	330	175	135	135	135	125	110	95	75	55	30	5
	E	190	60	340	560	680	695	630	505	335	135	135	125	110	95	75	55	30	5
	SE	185	15	140	315	475	580	625	615	545	420	260	125	110	95	75	55	30	5
	S	155	5	30	55	75	180	320	435	510	540	510	435	320	180	75	55	30	5
	SW	185	5	30	55	75	95	110	125	260	420	545	615	625	580	475	315	140	15
	W	190	5	30	55	75	95	110	125	135	135	335	505	630	695	680	560	340	60
	NW	140	5	30	55	75	95	110	125	135	135	135	175	330	460	530	510	360	75
	H	330	10	95	230	385	535	660	765	830	850	830	765	660	535	385	230	95	10
July 23 and May 21	N	75		135	155	85	90	105	120	130	135	130	120	105	90	85	155	135	
	NE	125		280	465	505	435	310	150	130	135	130	120	105	90	70	45	20	
	E	180		275	525	665	695	630	505	330	135	130	120	105	90	70	45	20	
	SE	190		120	305	480	595	645	635	565	440	280	120	105	90	70	45	20	
	S	165		20	45	70	200	345	465	540	570	540	465	345	200	70	45	20	
	SW	190		20	45	70	90	105	120	280	440	565	635	645	595	480	305	120	
	W	180		20	45	70	90	105	120	130	135	330	505	630	695	665	525	275	
	NW	125		20	45	70	90	105	120	130	135	130	150	310	435	505	465	280	
	H	305		65	195	350	500	630	730	795	820	795	730	630	500	350	195	65	
August 24 and April 20	N	50		5	75	55	75	95	110	115	120	115	110	95	75	55	75	5	
	NE	90		15	305	405	365	240	110	115	120	115	110	95	75	55	30	0	
	E	150		15	370	585	660	615	495	320	120	115	110	95	75	55	30	0	
	SE	190		10	240	460	610	685	680	615	490	325	135	95	75	55	30	0	
	S	185		0	30	95	250	405	530	615	640	615	530	405	250	95	30	0	
	SW	190		0	30	55	75	95	135	325	490	615	680	685	610	460	240	10	
	W	150		0	30	55	75	95	110	115	120	320	495	615	660	585	370	15	
	NW	90		0	30	55	75	95	110	115	120	115	110	240	365	405	305	15	
	H	245		0	95	240	395	530	635	700	725	700	635	530	395	240	95	0	
September 22 and March 22	N	30				30	55	70	85	95	100	95	85	70	55	30			
	NE	50				215	240	145	85	95	100	95	85	70	55	30			
	E	110				365	535	545	450	290	100	95	85	70	55	30			
	SE	170				315	545	670	695	640	525	360	180	70	55	30			
	S	200				100	270	445	580	670	700	670	580	445	270	100			
	SW	170				30	55	70	180	360	525	640	695	670	545	315			
	W	110				30	55	70	85	95	100	290	450	545	535	365			
	NW	50				30	55	70	85	95	100	95	85	145	240	215			
	H	165				95	230	365	470	540	560	540	470	365	230	95			
October 23 and February 20	N	20				0	25	45	65	70	75	70	65	45	25	0			
	NE	25				10	110	60	65	70	75	70	65	50	25	0			
	E	70				25	315	410	370	240	75	70	65	50	25	0			
	SE	135				20	350	545	625	600	505	360	190	50	25	0			
	S	180				10	195	390	550	650	680	650	550	390	195	10			
	SW	135				0	25	50	190	360	505	600	625	545	350	20			
	W	70				0	25	50	65	70	75	240	370	410	315	25			
	NW	25				0	25	50	65	70	75	70	65	60	110	10			
	H	95				0	90	200	300	365	385	365	300	200	90	0			
November 21 and January 21	N	10					5	25	40	50	55	50	40	25	5				
	NE	10					10	25	40	50	55	50	40	25	5				
	E	35					50	235	255	180	55	50	40	25	5				
	SE	95					60	335	470	490	425	305	165	40	5				
	S	130					35	255	430	545	580	545	430	255	35				
	SW	95					5	40	165	305	425	490	470	335	60				
	W	35					5	25	40	50	55	180	255	235	50				
	NW	10					5	25	40	50	55	50	40	25	10				
	H	50					5	80	155	215	235	215	155	80	5				
December 22	N	10						15	30	40	45	40	30	15					
	NE	10						15	30	40	45	40	30	15					
	E	25						150	205	150	45	40	30	15					
	SE	75						225	385	425	375	270	140	30					
	S	105						175	360	475	515	475	360	175					
	SW	75						30	140	270	375	425	385	225					
	W	25						15	30	40	45	150	205	150					
	NW	10						15	30	40	45	40	30	15					
	H	35						40	110	160	175	160	110	40					

Basis for tabulated values: 1. Direct Radiation factor (sky clarity), $k_c = 0.95$ 3. Ground Reflectance factor, $k_r = 0.2$
2. Cloudiness factor 0 4. Altitude = 0 to 300 m

Table 4.7 Total solar intensities on vertical and horizontal surfaces (W/m²) for latitude 51.7°N

	Thermal comfort standards					
	'Satisfactory'			'Minimum acceptable'		
Type of glass	clear			clear		
Type of shade	none	internal	external	none	internal	external
Cross-ventilation (i.e. doors on to corridor open)	0.04	0.05	0.04	0.10	0.27	0.40
Side ventilation (i.e. doors on to corridor closed)	–	–	0.35	–	0.23	0.40

Table 4.8 Maximum glass/floor area ratios for thermal comfort in summer in naturally ventilated offices

Table 4.9 sets out various combinations of glazing system and internal, or between glass, louvred blinds. When the louvres are closed, these can be considered as equivalent to roller blinds or curtains of similar performance category. The common criterion of window performance is the *shading coefficient*, expressed as a fraction of the total solar transmission through clear single glazing (87% of incident energy). The shading coefficient depends on the solar optical performance of the glass and the material of the blind.

The reflectance is the characteristic that influences the performance of a blind. A high performance blind has a reflectance of 0.50 or greater, medium performance of 0.40 and low performance of 0.30 or less.

The common arrangement of horizontal louvres set at 45° to the plane of the window has been used in establishing the figures in the table (Figure 4.11). Other criteria relate to the CIBS normal exposure grade, i.e. outside surface resistance – 0.055 m²°C/W, inside surface resistance – 0.123 m²°C/W, and in the case of double glazing with an air space of 12 mm a resistance of 0.154 m²°C/W, and with a 20 mm air space and internal blind a resistance of 0.167 m²°C/W.

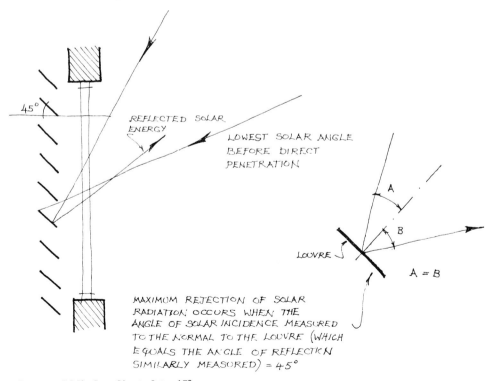

Figure 4.11 Louvred blinds adjusted to 45°

| Window design | Blind performance | Opaque louvre material | | | Translucent louvre material | | | Thermal fracture hazard | |
| | | Shading coefficient | | | Shading coefficient | | | | |
		Short wave	Long wave	Total	Short wave	Long wave	Total diff. °C	Temp. diff. °C	Remarks
Single glazed without blind									
CG	–	0.92	0.05	0.97	0.92	0.05	0.97	–	–
HAG	–	0.51	0.18	0.69	0.51	0.18	0.69	–	–
Single glazing with louvres closed									
CG	High	0	0.36	0.36	0.38	0.16	0.54	15	S
CG	Medium	0	0.50	0.50	0.38	0.25	0.63	–	–
CG	Low	0	0.63	0.63	0.38	0.34	0.72	13	S
HAG	High	0	0.40	0.40	0.21	0.27	0.48	32	S
HAG	Medium	0	0.46	0.46	0.21	0.31	0.52	–	–
HAG	Low	0	0.52	0.52	0.20	0.36	0.56	30	S
Single glazing with louvres at 45°									
CG	High	0.10	0.43	0.53	0.34	0.19	0.53	15	S
CG	Medium	0.07	0.55	0.62	0.31	0.31	0.62	–	–
CG	Low	0.05	0.65	0.70	0.28	0.42	0.70	13	S
HAG	High	0.05	0.41	0.46	0.17	0.29	0.46	32	S
HAG	Medium	0.03	0.47	0.50	0.15	0.35	0.50	–	–
HAG	Low	0.02	0.52	0.54	0.14	0.40	0.54	30	S
Double glazing without blind									
CG	–	0.74	0.10	0.84	0.74	0.10	0.84	–	–
HAG	–	0.41	0.14	0.55	0.41	0.14	0.55	–	–
Double glazing with louvres between, closed									
CG	High	0	0.15	0.15	0.32	0.10	0.42	28	S
CG	Medium	0	0.21	0.21	0.31	0.14	0.45	–	–
CG	Low	0	0.26	0.26	0.31	0.18	0.49	25	S
HAG	High	0	0.19	0.19	0.17	0.16	0.33	49	S in G
HAG	Medium	0	0.21	0.21	0.17	0.17	0.34	–	–
HAG	Low	0	0.23	0.23	0.17	0.19	0.36	46	S in G, B/M
Double glazing with louvres between, at 45°									
CG	High	0.09	0.19	0.28	0.28	0.12	0.40	28	S
CG	Medium	0.05	0.24	0.29	0.25	0.16	0.41	–	–
CG	Low	0.04	0.27	0.31	0.23	0.20	0.43	25	S
HAG	High	0.04	0.20	0.24	0.14	0.16	0.30	49	S
HAG	Medium	0.03	0.22	0.25	0.13	0.18	0.31	–	–
HAG	Low	0.02	0.23	0.25	0.11	0.20	0.31	46	S
Double glazing with louvres closed									
CG	High	0	0.40	0.40	0.31	0.22	0.53	20	S
CG	Medium	0	0.50	0.50	0.31	0.29	0.60	–	–
CG	Low	0	0.60	0.60	0.31	0.35	0.66	16	S
HAG	High	0	0.32	0.32	0.17	0.22	0.39	41	S/M, B/W
HAG	Medium	0	0.37	0.37	0.17	0.25	0.42	–	–
HAG	Low	0	0.42	0.42	0.17	0.28	0.45	37	S/W, B/C
Double glazed with internal louvres, at 45°									
CG	High	0.08	0.44	0.52	0.27	0.25	0.52	20	S
CG	Medium	0.05	0.53	0.58	0.25	0.33	0.58	–	–
CG	Low	0.04	0.60	0.64	0.22	0.42	0.64	16	S
HAG	High	0.04	0.32	0.36	0.14	0.22	0.36	41	S/M, B/W
HAG	Medium	0.03	0.37	0.40	0.13	0.27	0.40	–	–
HAG	Low	0.02	0.40	0.42	0.11	0.31	0.42	37	S/W, B/C

CG = clear glass, HAG = heat absorbing glass, S = safe, S/M = safe in metal
S in G = safe in gaskets, S/W = safe in wood, B/W = borderline in wood, B/C = borderline in concrete
B/M = borderline in metal

Table 4.9 Shading coefficients of various glazing systems in combination with louvres – the danger of thermal fracture

The short-wave component of the total shading coefficient relates to that portion of the incident radiation that is directly transmitted through the window with no change in wavelength. The long-wave component relates to that portion that is absorbed and then released inwards by conduction, convection or long-wave radiation.

The short-wave component (SCSW) can be calculated from the following:

$$SCSW = C \times T_g \times T_b$$

in which C = a constant (single glazing with blinds inside = 1.190,
double glazing with blinds between = 0.985,
double glazing with blinds inside = 0.980)
T_g = transmittance of glass (the outer leaf in double glazing)
T_b = transmittance of blind.

It should be borne in mind that only the short wavelength component of solar heat gain is subject to delaying and reducing by the thermal capacity of the structure (see Chapter 5).

Heat gain through vertical windows can be very easily established by using graphical aids, obtainable from the Environmental Advisory Service of Pilkington Brothers Ltd[4]. These consist of two graphs used in combination. Graph 1 is drawn for heat gain through single glazing; Graph 2 (on transparent film) is a sun path diagram for the relevant latitude (Figure 4.12).

Graph 2 is superimposed over Graph 1 with the lower axes coinciding. It is then slid along the axis until its arrow coincides with the appropriate orientation of the window. The instantaneous heat gain for any hour of the day and month of the year can then be read off Graph 1. This should next be modified by the haze factor from Table 4.5. The revised heat gain figure is then multiplied by the shading coefficient to establish the total short and long wavelength radiation through the glazing. The proportion of each wavelength component in relation to the particular glazing system is given on Table 4.9.

One final aspect of internal sun shading should be considered. Because high performance blinds reflect a large proportion of the transmitted energy back through the glazing for a second time, the absorbed energy in the glass is greatly increased and, therefore, its temperature increases. This can lead to excessive temperature differentials being built up between the centre of the pane and the shaded perimeter. Table 4.9 gives some of these temperature differences at peak radiation intensity of 750 W/m² and a diurnal range of outdoor temperature of 10°C. It also gives recommendations on frame types to use to avoid danger.

TRANSMISSION OF RADIANT HEAT THROUGH OPAQUE WALLS AND ROOFS

As explained earlier, in temperate climates with low levels of solar radiation this aspect of design can almost be ignored, particularly in the case of heavyweight structures. Nevertheless, opaque parts of the building shell react to solar radiation in a very similar way to glazed areas – heat is absorbed and, in time, part of it is transmitted to the inside of the room. It is, however, in respect of speed where they differ. Glass, as we have seen, has an almost instantaneous reaction to solar radiation, whereas a solid wall or roof has a mass which will absorb the heat and then gradually transmit it to the interior, depending on its transmittancy, or U value. This process may well take a long time and often the radiation peak passes before the process is complete. If this happens, some of the heat stored in the wall will slowly start to be transmitted outwards again.

In effect the wall becomes a sort of thermal storage block, slow to heat up and slow to lose its stored heat. A high thermal capacity structure can even out the peaks and troughs of temperature and help to stabilise the internal resultant temperature. This is a particularly fortunate property in those hot parts of the world which experience inland continental climates, combining hot days with relatively cold nights.

To calculate the heat transmission through walls and roofs it is first necessary to establish what precisely is to be considered as being the outside temperature. Just as

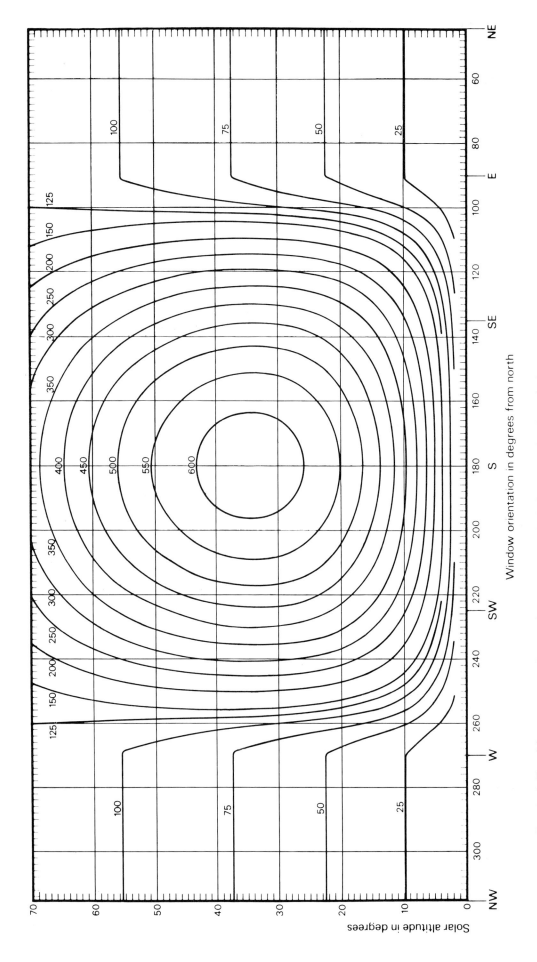

Figure 4.12(a)   Graph 1. Total instantaneous heat gain through vertical single glazing in W/m²

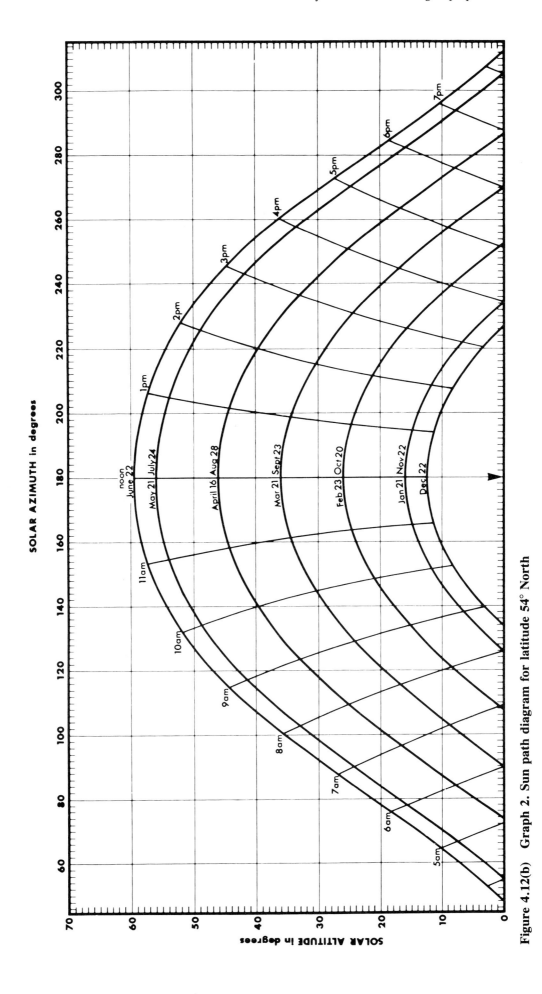

SOLAR AZIMUTH in degrees

SOLAR ALTITUDE in degrees

Figure 4.12(b)   Graph 2. Sun path diagram for latitude 54° North

5 The building shell: its insulation and thermal response

In the previous chapter some ways in which the building shell can help to control the thermal penetration of the internal environment by the external climate have been considered. Now, the part the building shell can play in preserving an unnatural internal climate will be examined. This is the 'overcoat' effect of the building shell, or the way in which it insulates its 'body' temperature from an external environment which can be either too hot, or too cold, for the comfort of the occupants of the internal space.

In temperate climates the problem is usually one of heat loss; in more tropical climates it can be one of heat gain. Undesirable heat loss or gain can also be experienced between adjacent internal areas in the same building. The discussion in this chapter will concern the movement of heat from the inside to the outside, or from the outside to the inside, through enclosing elements of structure. Whatever the direction of heat, the same principles largely apply.

First we will examine the way heat moves through a solid or semi-solid element of structure when the temperature difference on each side of the structure remains relatively static; then the same movement will be studied when the heat input, either internally or externally, is fluctuating. The object of both studies is to produce data, from which can be designed internal spaces which are comfortable to occupy, in spite of temperature swings or the special demands for heat brought about by intermittent occupancy.

SOURCES OF INTERNAL HEAT

The internal climate of the building is the result of intentional heat input – through the operation of the heating appliances – and unintentional casual heat gain from the occupants or equipment in the space. However unintentional casual gain may be, it is unavoidable if people are going to occupy the space; but it need not be a complete disadvantage. If its size is accurately predicted, it can be used to decrease the artificial heat input, thereby reducing fuel consumption and cost.

Some casual heat inputs are considerable. They need to be assessed before attempting to size the heating plant. In fact, in thermally well-designed buildings the casual heat input can largely remove the need for additional heating during a substantial part of the working day in the heating season.

It has been estimated that the energy used for lighting and running miscellaneous electrical equipment, such as radios and television sets, in the average, low rateable value house, produces 1000 MJ of space heating during the heating season. Cooking produces between 4000 and 6500 MJ, while an independent solid fuel boiler for domestic hot water contributes 50% of the calorific value of its fuel to space heating – gas and electric hot water heating produce rather less. The lowest contribution is

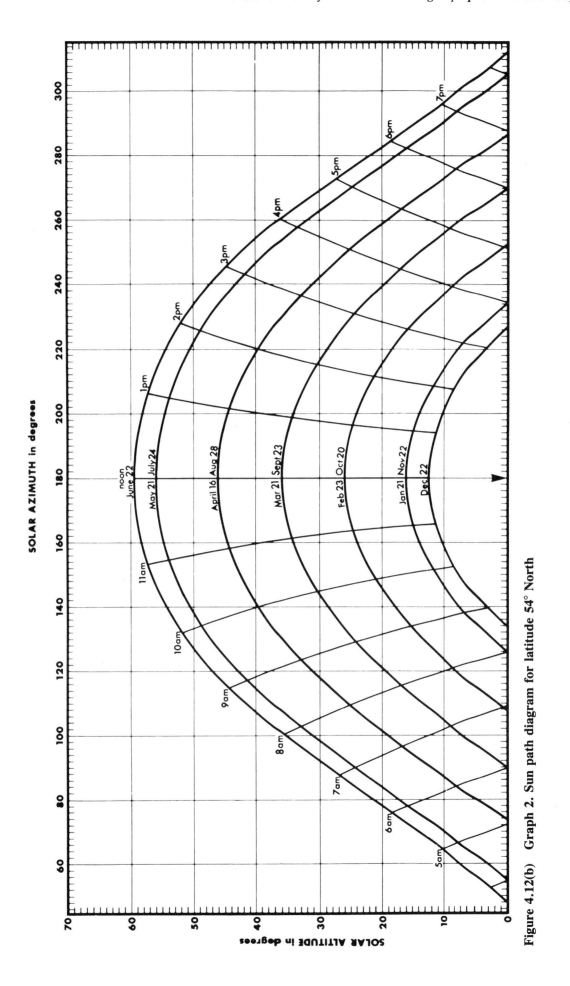

Figure 4.12(b) Graph 2. Sun path diagram for latitude 54° North

5 The building shell: its insulation and thermal response

In the previous chapter some ways in which the building shell can help to control the thermal penetration of the internal environment by the external climate have been considered. Now, the part the building shell can play in preserving an unnatural internal climate will be examined. This is the 'overcoat' effect of the building shell, or the way in which it insulates its 'body' temperature from an external environment which can be either too hot, or too cold, for the comfort of the occupants of the internal space.

In temperate climates the problem is usually one of heat loss; in more tropical climates it can be one of heat gain. Undesirable heat loss or gain can also be experienced between adjacent internal areas in the same building. The discussion in this chapter will concern the movement of heat from the inside to the outside, or from the outside to the inside, through enclosing elements of structure. Whatever the direction of heat, the same principles largely apply.

First we will examine the way heat moves through a solid or semi-solid element of structure when the temperature difference on each side of the structure remains relatively static; then the same movement will be studied when the heat input, either internally or externally, is fluctuating. The object of both studies is to produce data, from which can be designed internal spaces which are comfortable to occupy, in spite of temperature swings or the special demands for heat brought about by intermittent occupancy.

SOURCES OF INTERNAL HEAT

The internal climate of the building is the result of intentional heat input – through the operation of the heating appliances – and unintentional casual heat gain from the occupants or equipment in the space. However unintentional casual gain may be, it is unavoidable if people are going to occupy the space; but it need not be a complete disadvantage. If its size is accurately predicted, it can be used to decrease the artificial heat input, thereby reducing fuel consumption and cost.

Some casual heat inputs are considerable. They need to be assessed before attempting to size the heating plant. In fact, in thermally well-designed buildings the casual heat input can largely remove the need for additional heating during a substantial part of the working day in the heating season.

It has been estimated that the energy used for lighting and running miscellaneous electrical equipment, such as radios and television sets, in the average, low rateable value house, produces 1000 MJ of space heating during the heating season. Cooking produces between 4000 and 6500 MJ, while an independent solid fuel boiler for domestic hot water contributes 50% of the calorific value of its fuel to space heating – gas and electric hot water heating produce rather less. The lowest contribution is

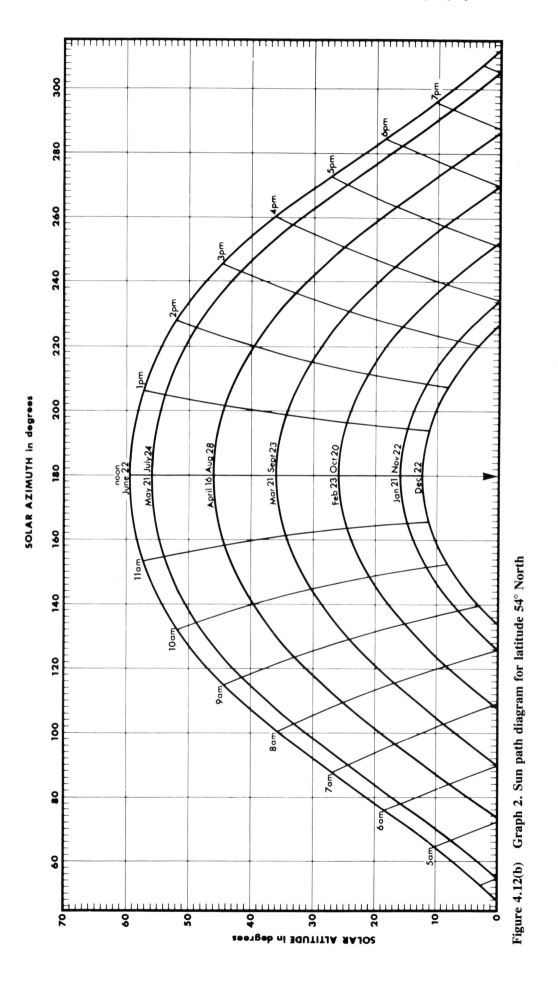

Figure 4.12(b)   Graph 2. Sun path diagram for latitude 54° North

in Chapter 2 the temperature of internal comfort was a combination of air and surface temperature, so external environmental temperature – called sol-air temperature – takes cognisance of solar radiation as well as air temperature. Once sol-air temperature is established the temperature difference between internal and external environments can be determined.

Material (clean)	Solar absorption coefficient (a)
Bricks:	
White glazed	0.25
Fletton light	0.4
Fletton dark	0.65
Stafford blue	0.9
White sand-lime	0.4–0.5
Red sand-lime	0.55–0.7
Stone:	
Limestone	0.3–0.5
White marble	0.45
Red granite	0.55
Roofs:	
Concrete tiles	0.65
Asphalt	0.9
Grey slates	0.8–0.9
Red tiles	0.4–0.8
Asbestos sheets (natural)	0.6
Galvanised iron	0.65
Lead sheeting	0.8
Mortar screed	0.8
Aluminium	0.2
Copper (tarnished)	0.65
Whitewashed roof or white tile	0.3–0.5
Water:	
1 m thick	0.56
2 m thick	0.61
3 m thick	0.64

When building surfaces become dirty, practical absorption coefficients can be derived using the tone of the clean material as the basis.

Surface of material when clean	Solar absorption coefficient (a)
Light	0.5
Medium	0.8
Dark	0.9

Table 4.10   Solar absorption coefficients for roofs and walls of different materials

Sol-air temperature can be defined as the outside air temperature which, in the absence of solar radiation, would give the same temperature distribution and rate of heat transfer through the structure as exists with the actual air temperature and incident solar radiation. The sol-air temperature for a *particular time* can be calculated from the following:

$$t_{eo} = t_{ao} + [R_{so} (a\ I_t - \epsilon I_L)]$$

in which t_{eo} = sol-air temperature (°C)

t_{ao} = outside air temperature (°C)

R_{so} = external surface resistance (m²°C/W)

a = absorption coefficient (see Table 4.10)

I_t = intensity of direct and diffuse solar radiation (W/m²)

ϵ = emissivity of outer surface to long-wave radiation (assumed 0.9 except for dull and polished aluminium where values are 0.2 and 0.5 respectively)

I_L = long-wave radiation from a black surface at air temperature (W/m²) – horizontal roof, 100 W/m²; vertical wall 0 W/m².

It should be remembered that this formula produces sol-air temperature for a particular moment. But sol-air temperature is a fluctuating thing, depending on climate, time of day, incidence of direct radiation, etc. Tables of average sol-air temperatures for the UK are given in the CIBS Guide.

A wall's or a roof's reaction to sol-air input is dependent on its material of construction, its colour and reflectancy and its thermal insulating properties. This is further complicated by the fact that solar radiation is not a static, but a cyclic input. All this has to be considered when assessing the way heat flows through external elements of construction. In the next chapter the flow of heat through such elements is considered together with the effects of cyclic input.

REFERENCES

1. Marsh, PH, *Air and rain penetration of buildings*. Construction Press, 1977.
2. Interdepartmental Construction Development Committee, *Technical Note 1: Performance requirements of windows*, HMSO.
3. Milbank, NO, *Energy consumption in 'other' buildings*, BRE, 1975.
4. Pilkington Brothers Ltd., *Solar heat gain through windows*, 1974.

5

The building shell: its insulation and thermal response

In the previous chapter some ways in which the building shell can help to control the thermal penetration of the internal environment by the external climate have been considered. Now, the part the building shell can play in preserving an unnatural internal climate will be examined. This is the 'overcoat' effect of the building shell, or the way in which it insulates its 'body' temperature from an external environment which can be either too hot, or too cold, for the comfort of the occupants of the internal space.

In temperate climates the problem is usually one of heat loss; in more tropical climates it can be one of heat gain. Undesirable heat loss or gain can also be experienced between adjacent internal areas in the same building. The discussion in this chapter will concern the movement of heat from the inside to the outside, or from the outside to the inside, through enclosing elements of structure. Whatever the direction of heat, the same principles largely apply.

First we will examine the way heat moves through a solid or semi-solid element of structure when the temperature difference on each side of the structure remains relatively static; then the same movement will be studied when the heat input, either internally or externally, is fluctuating. The object of both studies is to produce data, from which can be designed internal spaces which are comfortable to occupy, in spite of temperature swings or the special demands for heat brought about by intermittent occupancy.

SOURCES OF INTERNAL HEAT

The internal climate of the building is the result of intentional heat input – through the operation of the heating appliances – and unintentional casual heat gain from the occupants or equipment in the space. However unintentional casual gain may be, it is unavoidable if people are going to occupy the space; but it need not be a complete disadvantage. If its size is accurately predicted, it can be used to decrease the artificial heat input, thereby reducing fuel consumption and cost.

Some casual heat inputs are considerable. They need to be assessed before attempting to size the heating plant. In fact, in thermally well-designed buildings the casual heat input can largely remove the need for additional heating during a substantial part of the working day in the heating season.

It has been estimated that the energy used for lighting and running miscellaneous electrical equipment, such as radios and television sets, in the average, low rateable value house, produces 1000 MJ of space heating during the heating season. Cooking produces between 4000 and 6500 MJ, while an independent solid fuel boiler for domestic hot water contributes 50% of the calorific value of its fuel to space heating – gas and electric hot water heating produce rather less. The lowest contribution is

Activity	Type of building	Heat emission (W) at stated dry-bulb temperatures						
			15°		20°		24°	
		Total	*(s)*	*(l)*	*(s)*	*(l)*	*(s)*	*(l)*
Seated	Theatre, lounge	115	100	15	90	25	75	40
Light work	Office, restaurant	140	110	30	100	40	80	60
Walking, slow	Store, bank	160	120	40	110	50	85	75
Bench work	Factory	235	150	85	130	105	100	135
Medium work	Factory, dancing	265	160	105	140	125	105	160
Heavy work	Factory	440	220	220	190	250	135	305

The above figures are for an average male of body surface approximately 2m². Women can be taken as 85% and children 75% of these values.

(s) = sensible heat emission;   *(l)* = latent heat emission.

Table 5.1   Heat emission from the human body

from separate sink and bath heaters, but even these can contribute 200 MJ per week. Higher rateable value properties would presumably contribute higher casual heat inputs.

As for the occupants of the internal space, most of a person's food intake is emitted in the form of heat. Normal physiological processes liberate about 110 W from an average man – 80% of this is lost by convection and radiation, 20% by evaporation (mainly from the lungs). Females will emit about 85% of a male's output, children about 25%. Physical effort entails a greater heat output. For every watt of physical energy, the body produces 3 watts of heat – in other words, viewed as a machine, the human body is only 25% efficient.

Table 5.1 shows the levels of heat emission by the average male, during the course of various activities. Clearly, in establishing total casual heat gain from occupants of a space, allowance has to be made for the proportion of males, females and children likely to be in the space at any one time. Additionally, equipment of various types emits a certain amount of heat. Table 5.2 gives a few examples.

In temperate climates these casual heat gains have to be handled by natural ventilation, or air-conditioning cooling. In hot climates it is this heat, together with external heat gain through the structure, which needs to be adjusted by the air-conditioning equipment.

CAUSES OF HEAT LOSS FROM INTERNAL SPACES

The loss of heat from internal spaces can be either by ventilation loss or by fabric loss. Ventilation loss has already been mentioned briefly in Chapter 2 and again, with regard to air infiltration, in Chapter 4. It is sufficient here to note that the number of air changes per hour normally required for a particular type of accommodation will

Electrical equipment	Heat output (KW)	
	(s)	*(l)*
Grill – meat	1.2	0.6
Toaster	0.7	0.2
Coffee urn – 14 litre	0.7	0.5
Instrument washer and sterilizer	3.5	7.0
Sterilizer – 45 litre	1.2	4.8
Hair dryer	0.7	0.1

Table 5.2   Heat emission from some electrical equipment

have to be permitted by the building shell in a controlled manner, or by its equipment (if it has mechanical ventilation); but fortuitous ventilation above this amount should be prohibited by the detailing of the external wall and roof elements and the correct selection of windows and external doors. The heating equipment is required to cope with the replacement of the heat lost due to the design ventilation rates as stated in Table 2.2, Chapter 2; but should not need to be over-sized in order to cope with unintentional additional ventilation levels.

The chief contributions the building shell can make to the preservation of the internal heated environment are:

1. It should resist the passage of heat through its thickness to the outside air, or the reverse in tropical climates – in other words, it should provide thermal insulation.
2. It should respond to the fluctuations of ambient temperature, internally and externally, in such a way as to produce comfortable conditions for its occupants as soon as possible. This is in spite of intermittent demands on the building caused by living habits, or fluctuations of the external heat.

The building, Le Corbusier told us, is a machine for living in. Therefore, it must be a machine that responds to its occupants' needs. If it does not, its design is wrong.

THERMAL INSULATION: THE THEORY

All materials to a greater or less extent transmit heat. A lump of solid material strives by conduction – a process whereby neighbouring molecules of the material pass on temperature variations, one to the other – to produce temperature equilibrium. The part of the material closest to the heat source will convey that heat to neighbouring molecules and thence the heat will flow from molecule to molecule to the remotest part of the material. Not stopping at eventually achieving a temperature equilibrium within itself, the material will then proceed to radiate its heat to all colder adjacent objects, depending upon a property of the material called emissivity, and to lose its heat to the surrounding cooler air by convection or conductance.

This ability to conduct heat is to some extent a property of all materials. The rate at which a material will conduct heat depends on its conductivity (k) – this is a property of the material – and its thickness. Some materials have a more ready ability to conduct heat than others, an ability usually dictated by the structure of the material. The more dense and compact the structure, such as most metals and glass, the quicker the heat will be conducted. Those materials of a more 'open' structure, containing, maybe, imprisoned air bubbles, have a greater resistance to the passage of heat. Hence the best insulating materials are lightweight, usually of a cellular or fibrous nature (expanded polystyrene or mineral wool are good examples), but lacking in mechanical strength.

One type of material with insulating qualities which does not comply with this description is that which achieves its effect by reflecting radiant heat (for instance, aluminium foil-faced building paper). Because of this material's reflective surface, it tends not to absorb radiant heat. If used fronting an internal cavity in a composite building component, material with this property can lead to an increase in the thermal resistance of the cavity.

The thermally successful building envelope will discourage heat loss or gain by presenting the greatest resistance to the passage of heat through its thickness (inside to outside in cold climates; outside to inside in hot) that is consistent with the basic economics of the structure. Because good insulating material tends to lack strength, a single material shell is unlikely to provide both structural strength and adequate insulation. For instance, dense concrete thick enough to provide sufficient thermal insulation would be considerably more substantial than would be required for structural reasons – hence the use of this would be an uneconomic way of providing a thermally efficient shell. Therefore the use of composite enclosing components made up of several materials with differing characteristics, some of which are present for their strength or weather-protective qualities, some for their insulation, provides an economic method of solving the problem. Air spaces, or cavities, may be deliberately included in these composite components, because these form a barrier to conducted heat and, thereby, increase the thermal resistance of the component. Conduction only in the most limited form takes place in a cavity. Normal heat passage across narrow air spaces is by radiation (or very slightly by convection, particularly in the case of a horizontal cavity with heat passing from below). The efficiency of the radiation will depend on the emissivity of the material bordering the cavity. This surface resistance forms another resistance to heat transfer.

Insulation, included in the external component of a building, not only increases the thermal resistance of the component, it also assists in the control of the internal surface temperature – an important contribution to internal thermal comfort (see Chapter 2).

The position of the insulation in a composite enclosing heavyweight component has an important influence on the way the component responds to heat changes on either side of it. If the insulation is on the side adjacent to the heated climate, it will cause the component's surface temperature to respond more quickly to temperature changes – an important consideration in intermittently heated buildings – but it will also shield the mass of the component from the heat, causing the component as a whole to have a slow response to temperature change.

For instance, if the external temperature is 0°C and the internal temperature 16°C, with internal insulation the temperature drop through the structure from inside to outside will be steep, the majority of the component being at or near 0°C. If, on the other hand, there is insulation on the cold side of the component, the mass of the component will heat up, giving a slower response (and consequently a longer time lag before the internal surface reaches an acceptable temperature), but the component becomes a thermal store, capable of retaining heat which can be later re-radiated to the internal space when its air temperature falls. The thermal capacity of the component (its ability to store heat) is being used to even out ambient temperature fluctuations. The temperature gradient across the component in this case is completely different, with the majority of the component at a temperature much

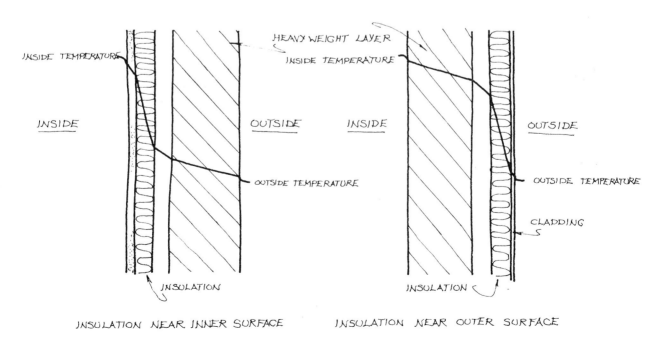

Figure 5.1 Temperature gradient through walls

closer to that of the heated space (Figure 5.1). This position of insulation is more appropriate to continuously heated buildings.

This question of insulation position in a component is not acknowledged in U value calculations. In these the overall transmittance of the component, air to air, is established and, as we shall see shortly in this chapter, the order of the individual elements in the component is immaterial. This is one of the limitations of the U value methods of assessing thermal performance. They are perfectly adequate if steady state conditions prevail and heat is flowing consistently. When fluctuations occur, such as are experienced in cyclic heating or solar radiation, the response of the structure is more complicated and more sophisticated techniques, therefore, have to be employed to establish precisely what is occurring. It was in response to this need that the *admittance* methods were developed, and these we shall be examining later in this chapter.

Here is another example of how the insulation position affects the way in which a structure performs thermally. In cloudless, hot climates experiencing considerable diurnal temperature fluctuations, the thermal capacity of the structure can be used to even out the highs and lows of temperature. Heat can be acquired slowly during the daytime and then transmitted to the interior at night. In these conditions, insulation placed near the inside surface of the building shell is advantageous in that it allows the solar heat input to be admitted to the structure. The mass, therefore, stores the daytime input, but due to the action of the insulation is slow to pass it on to the interior. This, to a certain extent, it will do at night, as well as re-radiating much of it outwards again to the clear night sky. In hot climates experiencing little diurnal temperature fluctuation there is some advantage in the mass of the component being protected from external heat gain, thereby allowing its thermal capacity to preserve the cool interior climate. It is, in these circumstances, normal to place the insulation on the outside of the component.

All the above is applicable only to heavyweight structures (brick, stone or concrete). In lightweight structures the position of the insulation is less critical owing to the structures' low thermal capacity.

The whole question of temperature fluctuation and admittance procedures will be returned to later in the chapter; but first the steady state of heat transfer should be understood.

THERMAL INSULATION – THE TERMS DEFINED

A building's thermal performance is based on the ability of its enclosing elements to conduct heat from one side of the shell to the other – their thermal transmittances. The thermal transmittance (U value) of an enclosing element is arrived at by considering the physical properties of the materials of which it is comprised. These properties are:

Thermal conductivity   Thermal conductivity (k) is a property of a homogeneous material and is a measure of its ability to conduct heat. It is the amount of heat (in watts) that a m² of the material will transmit in a unit time through a unit thickness (m) per °C of temperature difference. It is expressed in W/m°C.

Thermal resistivity   Thermal resistivity (r) is also a property of the material. It is the reciprocal of k, or 1/k, or m°C/W.

Thermal resistance   Thermal resistance (R) is the actual resistance of a particular thickness of a material and is established by multiplying a material's resistivity (r) by its thickness in metres. It is expressed as m²°C/W.

Thermal transmittance   Thermal transmittance (U) is a value for the performance of an element of structure, comprising one or more materials, of a given total thickness. It is a measure of its ability to transmit heat under steady state conditions. It is the quantity of heat that will pass through unit area of element, in unit time, per unit difference in temperature between inside and outside environments. It is an air-to-air value and takes into account the resistances of the inside and outside surfaces of the element to heat transfer. It also takes into account the resistance of any cavity or air space the element may contain. It is calculated as the reciprocal of the sum of the resistances of each layer of construction, the surface resistances and the cavity resistances. It is expressed as W/m²°C.

It should be noted that *thermal conductance* (C) may occasionally be encountered, expressed also in W/m²°C. Thermal conductance is the reciprocal of thermal resistance (R) and is a measure of the ability of the component to transmit heat from one surface to another. The difference between a U value and a C value is that a U value takes into consideration all surface resistances. It is a measure of heat transfer from air to air, not just surface to surface.

It will be understood from the above that the establishment of the U value of a structure will depend on knowing the conductivities (k values) of individual materials of the structure, the surface resistances and the cavity resistances.

STANDARD U VALUES

Until comparatively recently there has been confusion over the establishment of authoritative thermal transmittance values. Data had been deduced by a variety of means, sometimes by computation, sometimes by test. Not always were the same methods of test employed. At times air-dry (conditioned) specimens were tested in the laboratory; at others, materials were tested under natural exposure conditions.

Building element	Surface emissivity*	Heat flow	R_{si} (m²°C/W)
Walls	High	Horizontal	0.123
	Low	Horizontal	0.304
Ceilings or roofs,	High	Upwards	0.106
flat or pitched	Low	Upwards	0.218
Floors			
Ceilings and floors	High	Downwards	0.150
	Low	Downwards	0.562

* Emissivity should be taken as 'high' for all normal building materials except unpainted or untreated metallic surfaces such as aluminium or galvanised steel. These should be regarded as 'low'.

Table 5.3   Internal surface resistances (R_{si})

Building element	Surface emissivity*	R_{so} Surface resistance for stated exposure (m²°C/W)		
		Sheltered	Normal	Severe
Wall	High	0.08	0.055	0.03
	Low	0.11	0.067	0.03
Roof	High	0.07	0.045	0.02
	Low	0.09	0.053	0.02

* Emissivity should be taken as 'high' for all normal building materials except unpainted or untreated metallic surfaces such as aluminium or galvanised steel. These should be regarded as 'low'.

Table 5.4   External surface resistances (R_{so})

The resultant values were consequently inconsistent. At the same time doubts began to be expressed as to the effect of site exposure on external surface resistances and the validity of basing heat loss calculations on a temperature differential between air temperatures on each side of the structure. It became clear that some standard method was required whereby all materials were assessed against common criteria and then ways established of modifying these results for particular conditions in use.

After considerable work, the BRS (as it was then called) produced a system of *standard* U values, which were explained in a current paper (CP79/68) prior to their introduction in the (then) IHVE Guide Book 1970. The object of the new standard procedure was to arrive at a series of 'normal' U values, typical of conditions likely to obtain in practice, with methods of modification to accommodate different conditions. From that time, U values have been arrived at by computation, using standard values for surface resistance, moisture content and related thermal resistances, linked by standardised calculation methods.

The bases of this procedure were:

1. Thermal conductivity of porous elements was based on appropriate moisture contents, depending on whether the element in question was protected from or exposed to the weather (or condensation).
2. External surface resistances were expressed in accordance with the exposure of the element under consideration, depending on geographical location and/or height, but without consideration for orientation (see Chapter 3).
3. The temperature difference across the element was no longer the difference between internal and external *air* temperature, but was the difference between internal comfort temperature (Chapter 2) and sol-air temperature (Chapter 4).

Standard thermal transmittances can be used in heat loss calculations, assuming continuous heating, sunless conditions and steady temperatures. Modified external surface resistances have to be used in sheltered or severe exposures, and the moisture contents of the materials comprising the element, if different from standard, may have to be allowed for.

Type of air space		R_{cav} thermal resistance including internal surfaces (m²°C/W)	
Thickness (mm)	Surface emissivity	Heat flow horizontal or upwards	Heat flow downwards
5	High	0.11	0.11
	Low	0.18	0.18
20 or more	High	0.18	0.21
	Low	0.35	1.06
	High emissivity planes and corrugated sheets in contact	0.09	0.11
	Low emissivity multiple foil insulated with airspace on one side	0.62	1.76

Applicable to roof spaces and cavity wall cavities in spite of the small amount of ventilation present.

Table 5.5 Cavity resistances for unventilated cavities (R_{cav})

CALCULATING THE U VALUE

In a simple structure, in which there is no heat bridging, the U values for walls and roof can be obtained from:

$$U = \frac{1}{R_{si} + R_{so} + R_{cav} + R_1 + R_2 \dots}$$

in which R_{si} = internal surface resistance (see Table 5.3)
R_{so} = external surface resistance (see Table 5.4)
R_{cav} = resistance of any cavity within the element
R_1, R_2 = resistance of individual materials comprising the element.

Clearly, the greater the resistances in this equation, the smaller the U value – and the more thermally efficient, from an insulation point of view, is the element being considered.

Resistances of air spaces   The resistance of air spaces within cavity construction depends on the following factors:

1. The thickness of the air space. Resistance increases with width up to a maximum at 20 mm.
2. Surface emissivity. Most common building materials have high emissivity, resulting in radiation producing two-thirds of the heat transfer across an air space. Materials with low emissivity have bright surfaces, such as aluminium foil.
3. The plane of the air space and the direction of heat flow. A horizontal air space offers greater resistance to downward than upward heat transfer, convection currents assisting upward transfer. Horizontal heat transfer in a vertical cavity is the same as upward flow in a horizontal cavity.
4. The degree of ventilation. Ventilation obviously provides an alternative route for heat transfer, but the conditions experienced can be very variable. Ventilation may be deliberate, or fortuitous.

Tables 5.5 and 5.6 indicate standard thermal resistances for unventilated and ventilated air spaces. It should be noted that the normal degree of ventilation installed in roof spaces to avoid condensation and similar ventilation in cavity walls, does not for this purpose constitute ventilated air spaces.

Type of air space (minimum thickness 20 mm)	R_{cav} thermal resistance ($m^2 °C/W$)
Between asbestos-cement or black painted metal cladding with unsealed joints and high emissivity lining	0.16
As above with low emissivity surface facing the airspace	0.30
Loft space between flat ceiling and unsealed asbestos-cement or black metal cladding pitched roof	0.14
As above with aluminium cladding instead of black metal cladding, or with a low emissivity upper surface on the ceiling	0.25
Loft space between flat ceiling and unsealed tiled pitched roof	0.11
Between tiles and roofing felt or building paper on pitched roof	0.12
Behind tiles on tile-hung wall	0.12

Table 5.6   Cavity resistances for ventilated cavities (R_{cav})

Material	Density kg/m³	k W/m°C
Asbestos-cement sheet	1360	0.25
	1600	0.40
	2000	0.55
Asbestos fibre (sprayed)	120–240	0.040–0.075
Asbestos insulation board	720	0.11
	800	0.14
Asphalt roofing	1600–2325	0.43–1.15
Bitumen roofing felt	960	0.19
Building paper	–	0.065
Carpeting	160–270	0.045–0.065
Cork granules	115	0.052
Corkboard	145	0.042
Fibre insulation board	240	0.053
	300	0.057
(bitumen impreg.)	430	0.070
Glass (sheet)	2500	1.05
(heat resisting)	2250	1.10
Glass wool, mat or quilt	25	0.040
loose wool blanket	145	0.042
rigid pipe sections	160	0.042
Hardboard, medium	600	0.08
standard	900	0.13
Metals:		
aluminium alloy (average)	2800	160
copper	8900	200
steel, carbon	7800	50
high alloy	8000	15
Mineral wool, felted	50	0.039
semi-rigid felted mat	130	0.036
loose felted slab or mat	180	0.042
rigid slab	155	0.050
Perlite, loose granules	65	0.042
Plaster, gypsum	1280	0.46
sand/cement	1570	0.53
vermiculite	640	0.187
perlite	600	0.19
Plasterboard, gypsum	950	0.16
perlite	800	0.18
Plastics, cellular		
expanded polystyrene	15	0.033
polyurethane foam (aged)	30	0.026
urea formaldehyde (cavity fill)	8–16	0.030–0.036
Plastics, solid		
epoxy glass fibre	1500	0.23
polycarbonate	1150	0.23
polyethylene, low density	920	0.35
high density	960	0.50
polystyrene	1050	0.17
Sarking felt	1100	0.20
Stone:		
Artificial – dry	1750	1.3
Granite– dry	2600	2.3
Limestone – dry	2180	1.5
Marble – dry	2500	2.0
Sandstone – dry	2000	1.3
Slate – dry	2700	1.9
Strawslab, compressed	330	0.098
Tiles:		
Clay	1900	0.85
Concrete	2100	1.10
Timber, across the grain:		
softwood	–	0.13
hardwood	–	0.15
plywood	530	0.14
Vermiculite, loose granules	100	0.065
Wood chipboard	800	0.15
Wood wool slabs	600	0.110

Table 5.7 Thermal conductivities (k) of some common building materials

Resistance of materials   The thermal resistance of the materials making up the building element can be obtained from the reciprocal of the material's k value, multiplied by its thickness in metres, as follows:

$$R = \frac{1}{k} \times t$$

A short list of k values for some typical building materials is given in Table 5.7. A very much more extensive list can be found in the CIBS Guide. These k values are for air-dry samples.

Materials such as those used in masonry walling – bricks, concrete blocks etc – are to some degree porous and are often used in positions where they are liable to become damp – for instance, in the outer leaf of a cavity wall. As thermal transmittance increases in proportion to the increase in moisture content it is clearly necessary to make an allowance for this fact in establishing a material's true, in-use k value.

It has been established that a standard value for moisture content of brickwork protected from the rain is 1%, and exposed to rain, 5%. Concrete's moisture content, protected, is 3% and exposed, 5%. As there is a substantial variation in thermal conductivity with the density of this type of material, a table of average k values has been produced (for both wet and dry conditions) applicable to a range of typical dry densities of blocks and bricks (see Table 5.8). If, however, the measured k value can be obtained for a particular material, this should always be used in preference in U value calculations. The test to establish the k value should have been carried out on specimens at a fairly low moisture content and should be adjusted to the appropriate standard moisture content in use, by the factors in Table 5.9.

Bulk dry density kg/m³	Thermal conductivity (W/m °C)		
	Brickwork protected from rain (1% moisture content)*	Concrete protected from rain (3% moisture content)*	Brickwork or concrete exposed to rain (5% moisture content)*
200	0.09	0.11	0.12
400	0.12	0.15	0.16
600	0.15	0.19	0.20
800	0.19	0.23	0.26
1000	0.24	0.30	0.33
1200	0.31	0.38	0.42
1400	0.42	0.51	0.57
1600	0.54	0.66	0.73
1800	0.71	0.87	0.96
2000	0.92	1.13	1.24
2200	1.18	1.45	1.60
2400	1.49	1.83	2.00

* Moisture content expressed as a percentage by volume.

Table 5.8   Thermal conductivity of masonry materials related to moisture content

Moisture content (% by volume)	1	3	5	10	15	20	25
Moisture factor	1.3	1.6	1.75	2.1	2.35	2.55	2.75

Table 5.9   Moisture factors for use with Table 5.8

HEAT BRIDGING AND DISCONTINUOUS STRUCTURES

Establishing the thermal transmittance of an element bridged by material of a higher or lower thermal conductivity is a more complex matter. If the bridging through the element is complete, that is the bridge passes completely through except for a thin cladding material on both faces, the U value of the whole element can be calculated by the following method:

1. First, assess the U value of the main construction between bridges.
2. Then, similarly, assess the U value of the structure at the bridge.
3. Then combine the two U values in proportion to their relative areas.

$$U = F_1U_1 + F_2U_2$$

in which U = the combined value of the element
U_1 = the U value of the construction between the bridges of total area F_1
U_2 = the U value of the construction at the bridge of total area F_2

A typical example of such a construction is a timber framed wall panel. A calculation of a U value for such a construction is given later in Example 2.

This method can be used when assessing the U values of walls of hollow blocks, provided that the air spaces are not less than 20 mm thick and are wide in proportion to their thickness. The method should not, however, be applied to multi-perforated bricks or blocks whose cavities do not comply with the above conditions.

Where elements are crossed by highly conductive heat bridges, such as metal members of symmetrical section (eg an RSJ or Universal column), the above method can be applied. The effective width of the bridge is taken as being the width of the flange in contact with the rear and front faces. The method cannot be used, however, when metallic bridges are in contact with metallic claddings or linings.

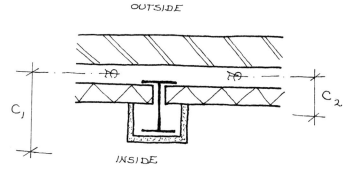

Figure 5.2   Bridging of the inner leaf

If only the inner leaf of a multi-layer construction is bridged (see Figure 5.2) the thermal transmittance is calculated as follows:

1. Calculate the conductance of the bridged and unbridged portions from the building's interior to the centre of the air space. The conductance of the bridged portion (C_1) is found by adding the inside surface resistance (R_{si}), the resistance of the bridged portion (R_1), half the air space resistance ($\frac{1}{2}R_{cav}$) and taking the reciprocal:

$$C_1 = 1 / (R_{si} + R_1 + \tfrac{1}{2}R_{cav})$$

The conductance of the unbridged portion is found similarly:

$$C_2 = 1 / (R_{si} + R_2 = \tfrac{1}{2}R_{cav})$$

Note: R_2 = the resistance of the unbridged portion.

2. Combine these conductances in proportion to their areas and the resistance of the inside half of the wall (R_{ic}) by taking the reciprocal of the combined conductance.

3. Calculate the resistance of the front half of the wall (R$_{co}$) by adding the remaining thermal resistances:

$$R_{co} = \tfrac{1}{2}R_{cav} + R_3 + R_{so}$$

4. Calculate the thermal transmittance by adding the two halves of the thermal resistances together and taking the reciprocal:

$$U = 1 \,/\, (R_{ic} + R_{co})$$

This method is applicable also to elements where only the outside leaf is bridged, or where either the inner or outer leaf of the construction is built of hollow blocks with cavities greater than 20 mm and wide in proportion to their thickness.

TYPICAL CALCULATIONS TO ESTABLISH U VALUES

EXAMPLE 1

Determine the U value of a cavity wall constructed of an external leaf of 1600 kg/m³ brickwork, 105 mm thick, 5% moisture content; a cavity 50 mm wide; an inner leaf of 750 kg/m³ aerated concrete blocks, 100 mm thick, 3% moisture content and 13 mm lightweight plaster (Figure 5.3) Exposure: normal.

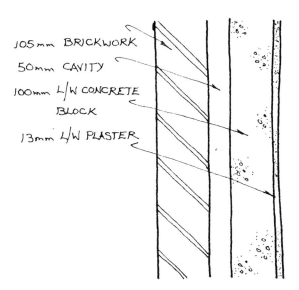

Figure 5.3 Example 1

k values of materials:

1. 1600 kg/m³ brickwork at 5% moisture content   = 0.73
2. 600 kg/m³ concrete blocks at 3% moisture content = 0.19
3. lightweight plaster, 13 mm thick   = 0.19

Material resistances:

brickwork $\dfrac{1}{0.73} \times 0.105$ = 0.144

concrete block $\dfrac{1}{0.19} \times 0.1$ = 0.526

plaster $\dfrac{1}{0.19} \times 0.013$ = 0.068

Total material resistance   = 0.738

Non-material resistances:

internal	= 0.123
external	= 0.055
cavity	= 0.180
Total resistance	1.096

$$U = \frac{1}{1.096} = 0.912 \text{ W/m}^2\,^\circ\text{C}$$

EXAMPLE 2

Determine the U value of a timber framed wall panel, 2.350 m × 600 mm, constructed of an external cladding of 13 mm bitumen impregnated fibreboard, 90 mm × 40 mm framing in softwood and 13 mm foil-backed plasterboard (Figure 5.4).

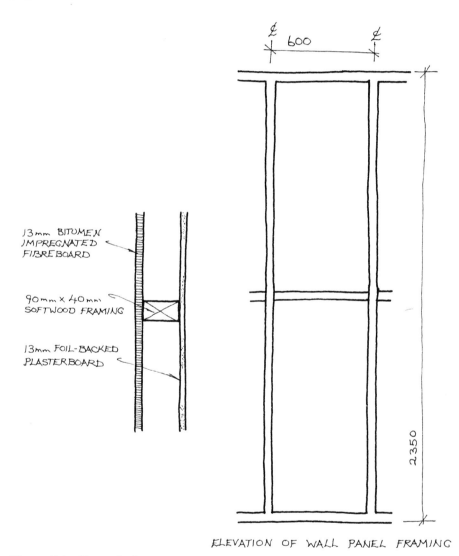

ELEVATION OF WALL PANEL FRAMING

Figure 5.4   Example 2

First determine U_1 (the U value between framing members)

k values of materials:

1. fibreboard   = 0.070
2. plasterboard = 0.160

Material resistances:

fibreboard $\dfrac{1}{0.070} \times 0.013 = 0.186$

plasterboard $\dfrac{1}{0.160} \times 0.013 = 0.081$

Total material resistance   0.267

Non-material resistances:

internal	= 0.123
external	= 0.055
cavity	= 0.620

Total resistance   1.065

$$U_1 = \frac{1}{1.065} = 0.94 \text{ W/m}^2\text{°C}$$

Then determine U_2 (the U value at the framing members):

k values of materials:

1. fibreboard   = 0.070
2. plasterboard = 0.160
3. studwork   = 0.130

Material resistances:

fibreboard $\dfrac{1}{0.070} \times 0.013 = 0.186$

plasterboard $\dfrac{1}{0.160} \times 0.013 = 0.081$

studwork $\dfrac{1}{0.130} \times 0.090 = 0.692$

Total material resistance   0.959

Non-material resistances:

internal = 0.123
external = 0.055

Total resistance   1.137

$$U_2 = \frac{1}{1.137} = 0.88 \text{ W/m}^2\text{°C.}$$

Determine the overall U value.

The total area of the wall panel is 1.41 m².

The area of framing (F_2)

$$= (2.35 \times 0.02 \times 2) + (0.56 \times 0.04 \times 2) + (0.56 \times 0.04)$$
$$= 0.161 \text{ m}^2$$

Area F_1 = total area − 0.161 = 1.41 − 0.161 = 1.249 m²

$$U = \frac{F_1 U_1 + F_2 U_2}{1.41} = \frac{(1.249 \times 0.94) + (0.161 \times 0.88)}{1.41}$$

$$= \frac{1.174 + 0.145}{1.41} = 0.93 \text{ W/m}^2\text{°C.}$$

Clearly, in this example the bridging effect of the timber framing members is negligible, as the thermal resistance of the stud and that of the cavity and foil-backing to the plasterboard are roughly the same. However, had the cavity been filled with insulation the effect would have been more marked, as we will show later.

The CIBS Guide lists an extensive series of U values for various constructions. A smaller list is given in Table 5.10. In using these, it should always be borne in mind that the same conditions must apply on site as those laid down in the table. Any alteration could have significant effect on the thermal performance of the building. In other words, the correct exposure grade must be established, the materials must be used at the stated moisture contents and the materials of construction must be precisely those stated. Otherwise it would be wise either to recalculate the U values afresh using the correct data, or to modify the stated U values to take account of any alteration.

METHOD OF ADJUSTING STATED U VALUES

The method to be employed is as follows:

1. Find the reciprocal of the U value. This is then the total resistance of the construction.
2. The resistances of any layers of the construction which are to be omitted, or modified in the proposed construction are then deducted from the total resistance.
3. The resistances of any new layers, or modified layers are then added to the total resistance.
4. The reciprocal of this new total resistance is then the revised U value of the modified construction.

EXAMPLE 3

Taking the construction specified in Example 1, examine the effect on the U value if 50 mm of loose blown mineral wool were installed in the cavity.

Existing U value = 0.912 W/m²°C.

$$\text{Total resistance} = \frac{1}{0.912} = 1.096$$

Mineral wool insulation: k value = 0.042

$$\text{Thermal resistance} = \frac{1}{0.042} \times 0.05 = 1.190$$

From total resistance above deduct the resistance of the cavity

$$1.096 - 0.18 = 0.916$$

Add mineral wool resistance $\underline{1.190}$

$$2.106$$

$$\text{Revised U value} = \frac{1}{2.106} = 0.47 \text{ W/m}^2{}°C.$$

Table 5.10 Thermal characteristics of various constructions

Construction	Thickness mm	Density kg	Vapour check or barrier	Steady state conditions (normal conditions)			Cyclic conditions (where figures are available)		
				R m²°C/W	U W/m²°C	Surface temp. difference °C (internal 20°C external 0°C)	Admittance Y W/m²°C	Decrement f	Surface F
Walls									
(a) brickwork	220	1700	none	0.467	2.14	5.26	4.4	0.49	0.53
dense plaster	13	1280							
(b) as (a) plus rendering	13	1280	none	0.495	2.02	4.96			
(c) as (a) but lower density brick	220	1600	none	0.505	1.98	4.86			
(d) as (a) but light-weight plaster	13	600	none	0.510	1.96	4.83	3.4	0.45	0.62
(e) as (b) but light-weight plaster	13	600	none	0.538	1.86	4.58			
(f) as (c) but light-weight plaster	13	600	none	0.546	1.83	4.50			
(a) concrete	150	2100	none	0.763	1.31	3.22	2.3	0.50	0.80
wood wool	50	600							
dense plaster	13	1280							
(b) as (a) plus rendering	13	1280	none	0.794	1.26	3.10			
(c) as (b) but light-weight plaster	13	600	none	0.833	1.20	2.94			
(a) tile hanging	150	750	none	1.053	0.95	2.34			
lightweight concrete		600							
lightweight plaster	13								

ALTERNATIVE b + e

ALTERNATIVE b + c

(continued overleaf)

69

Table 5.10 **Thermal characteristics of various constructions** (*continued*)

Construction	Thickness mm	Density kg	Vapour check or barrier	Steady state conditions (normal conditions)			Cyclic conditions (where figures are available)		
				R m²°C/W	U W/m²°C	Surface temp. difference °C (internal 20°C external 0°C)	Admittance Y W/m²°C	Decrement f	Surface F
(a) precast sandwich panel dense concrete polystyrene lightweight concrete	75 25 150	2100 15 1200	none	1.383	0.72	1.78	3.8	0.28	0.62
(a) brickwork – both leaves dense plaster	105 13	1700 1280	none	0.678	1.47	3.63	4.3	0.43	0.57
(b) as (a) but lightweight plaster	13	600	none	0.719	1.39	3.42	3.3	0.39	0.64
(c) as (a) but lower density brick	105	1600	none	0.724	1.38	3.40			
(d) as (c) but lightweight plaster	13	600	none	0.763	1.31	3.22			

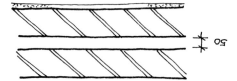

Construction	thickness (mm)	density	vapour check						
(a) brickwork lightweight concrete block dense plaster	105 100 13	1700 600 1280	none	1.036	0.96	2.37	2.9	0.56	0.77
(b) as (a) but lightweight plaster	13	600	none	1.077	0.93	2.28			
(c) as (b) but lower density brick	13	600	none	1.096	0.91	2.24			
(d) as (a) but with polystyrene in cavity	13	15	none	1.430	0.70	1.72	3.0	0.49	0.77
(e) as (c) but with polystyrene in cavity	13	15	none	1.490	0.67	1.65			
(f) as (c) but with mineral wool in cavity	50	180	none	2.106	0.47	1.17			
(g) as (c) but with urea formaldehyde foam in cavity and render	50 13	12 1280	none	2.604	0.38	0.94			
(a) lightweight concrete – both leaves render dense plaster	75 & 100 13 13	600 1280 1280	none	1.335	0.75	1.84	3.2	0.54	0.73
(a) brickwork – both leaves polystyrene/plasterboard laminate – with vapourcheck	105 25 (13–13)	1600	required	1.250	0.80	1.97			

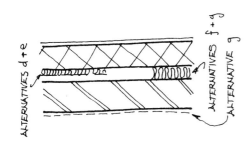

ALTERNATIVES d + e
ALTERNATIVES f + g
ALTERNATIVE g

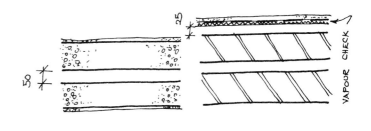

50
25
VAPOUR CHECK

(continued overleaf)

Table 5.10 Thermal characteristics of various constructions *(continued)*

Construction	Thickness mm	Density kg	Vapour check or barrier	Steady state conditions (normal conditions)			Cyclic conditions (where figures are available)		
				R m²°C/W	U W/m²°C	Surface temp. difference °C (internal 20°C external 0°C)	Admittance Y W/m²°C	Decrement f	Surface F
(a) brickwork	105	1600	required	2.174	0.46	1.13			
bitumen/fibreboard on	13	430							
timber framing									
mineral wool	50	130							
plasterboard	13	950							
Roofs – flat									
(a) asphalt	15	1700	ventilation under roofing	0.526	1.9	4.68	5.1	0.36	0.44
lightweight screed	75	600							
concrete	150	2100							
dense plaster	13	1280							
(b) asphalt	19	1700	ventilation as above	1.163	0.86	2.12	2.5	0.78	0.81
aerated concrete	150	500							
dense plaster	13	1280							
(c) polystyrene (closed cell)	50	15	ventilation as above	1.896	0.53	1.30			
bitumen felt	19	926							
concrete	150	2100							
dense plaster	13	1280							

Construction	Thickness (mm)	Density	Vapour check			
						1.0
						0.99
						0.87
(d) bitumen felt polystyrene metal decking	19 25	926 15	required	0.971	1.03	2.53
(e) as (d) plus ceiling of plasterboard	13	950	required	1.402	0.713	1.75

Roofs – pitched

Construction	Thickness (mm)	Density	Vapour check			
(a) tiles on battens roofing felt plasterboard	– – 13	1900 960 950	required	0.667	1.50	3.18
(b) as (a) plus sarking boarding	13	650	required	0.769	1.30	2.76
(c) as (b) plus glass fibre insulation	50	25	required	2.000	0.50	1.06

Construction	Thickness (mm)	Density	Vapour check			
(a) asbestos-cement sheet	–	1500	–	0.164	6.1	12.92
(b) as (a) plus lining of plasterboard	13	950	required	0.526	1.9	4.03
(c) as (b) plus mineral wool	60	180	required	1.954	0.51	1.08

VAPOUR CHECK — ALTERNATIVE (e)

BOARDING INSERTED HERE ALTERNATIVE b.

VAPOUR CHECK — ALTERNATIVE c

VAPOUR CHECK — ALTERNATIVES b + c — ALTERNATIVE c

(continued overleaf)

73

Table 5.10 Thermal characteristics of various constructions (*continued*)

Construction	Thickness mm	Density kg	Vapour check or barrier	Steady state conditions (normal conditions)			Cyclic conditions (where figures are available)		
				R m²°C/W	U W/m²°C	Surface temp. difference °C (internal 20°C external 0°C)	Admittance Y W/m²°C	Decrement f	Surface F
(a) corrugated aluminium	–	–	–	0.26	3.80	16.77			
(b) as (a) plus lining of plasterboard	13	950	required	0.526	1.90	4.03			
(c) as (b) plus mineral wool	60	180	required	1.954	0.51	1.08			

Glazing systems

	U values without frames
Single	5.6
Double	
20 mm air space	2.9
12 mm air space	3.0
6 mm air space	3.4
3 mm air space	4.0
Triple	
20 mm air space	2.0
12 mm air space	2.1
6 mm air space	2.5
3 mm air space	3.0
Roof glazing	6.6
Laylight with skylight	
ventilated	3.8
unventilated	3.0

74

EXAMPLE 4

Taking the wall panel in Example 2, examine the result of placing 50 mm of mineral wool insulation between the timber framing and adding a 105 mm skin of brickwork (1600 kg/m³, 5% moisture content) to the exterior (Figure 5.5).

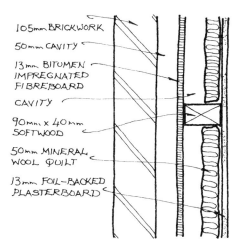

105mm BRICKWORK
50mm CAVITY
13mm BITUMEN IMPREGNATED FIBREBOARD
CAVITY
90mm x 40mm SOFTWOOD
50mm MINERAL WOOL QUILT
13mm FOIL-BACKED PLASTERBOARD

Figure 5.5 Example 4

Establish the revised U value for U_1 (U_2 remains unchanged)

Mineral wool quilt k value = 0.036 (Note; it is important to establish the k value of the particular form of mineral wool used)

Thermal resistance of 50 mm thickness $= \dfrac{1}{0.036} \times 0.05 = 1.389$

$U_1 = 0.94 = \dfrac{1}{1.065}$

Total resistance is therefore	1.065
Deduct cavity resistance	0.620
	0.445

Add cavity without low emissivity face	0.180
	0.625
Add insulation	1.389
	2.014

$U_1 = \dfrac{1}{2.014} = 0.49$ W/m²°C.

Note the significant difference between the U value through the stud and the U value between the studs now that insulation is added.

The overall U value has now become:

$$\frac{(1.249 \times 0.49) + (0.161 \times 0.88)}{1.41} = \frac{0.612 + 0.142}{1.41} = 0.534 \text{ W/m}^2{}^\circ\text{C}$$

Add the thermal resistance of the brick skin and the additional cavity.

Brickwork k = 0.73	Resistance	=	0.144 (see Example 1)
	Cavity resistance	=	0.180
	Total additional resistance		0.324

Add to the revised U_1 resistance: $2.014 + 0.324 = 2.338$
Add to U_2 resistance: $1.137 + 0.324 = 1.461$

Therefore $U_1 = 0.43$ and $U_2 = 0.68$

Overall $U = \dfrac{(1.249 \times 0.43) + (0.161 \times 0.68)}{1.41} = 0.46$ W/m²°C.

U VALUES

Ground floors next to earth

U values for unheated ground floors next to earth are given in Table 5.11. Clearly the ratio between the exposed perimeter of the floor and the area of the floor will affect its overall thermal transmission – the greatest heat loss being restricted to a relatively narrow zone at the outside edge of the floor. Transmission is either direct to air, or through the ground adjacent to the external air; that is, ground not heated up by the building above (Figure 5.6). In tropical climates, the heat flow is clearly in the opposite direction, but the principle remains unaltered. The significance of the shape of the floor is reflected in Table 5.11, in which U values are given for a range of ground floor dimensions with either four or two adjacent sides exposed to the exterior.

Dimensions of floor (m)	U values	
	Four exposed edges (W/m²°C)	Two exposed edges at right angles (Wm²°C)
Very long × 30	0.16*	0.09
× 15	0.28*	0.16
× 7.5	0.48*	0.28
150× 60	0.11	0.06
× 30	0.18	0.10
60× 60	0.15	0.08
× 30	0.21	0.12
× 15	0.32	0.18
30× 30	0.26	0.15
× 15	0.36	0.21
× 7.5	0.55	0.32
15× 15	0.45	0.26
× 7.5	0.62	0.36
7.5× 7.5	0.76	0.45
3× 3	1.47	1.07

* applies also to any floor of this breadth and losing heat from two parallel edges (breadth = distance between exposed edges).

Table 5.11 U values for solid floors next to earth

The U values stated are applicable to dense concrete floor slabs, with or without a bed of hardcore. The thickness of the slab is immaterial to the U value, as the transmittances of the slab and the earth are roughly the same. Hard, dense floor finishes, such as granolithic or clay tiling, or thin finishes of little insulation value, such as thermoplastic or vinyl tiles, will not affect the thermal performance of the floor.

In applying these U values, the full temperature difference between internal and external environmental temperatures must be used (t_c and t_{eo}; see Chapters 2 and 4) in spite of the fact that the underside temperature of the slab (particularly away from the building perimeter) will be higher than the outdoor temperature. An adjustment has been made in the U values to take account of this apparent discrepancy.

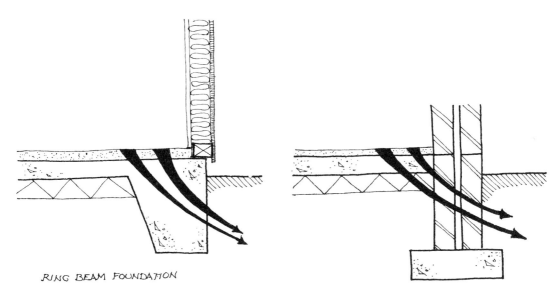

RING BEAM FOUNDATION

TRADITIONAL FOOTINGS

Figure 5.6   Heat loss through ground floors

The effect of moisture content in the floor slab (unlike moisture in walling materials) can be ignored.

Insulated solid ground floors

If a floor finish or screed with thermal insulating properties is used, the basic U value for that floor can be modified following the method set out above. An overall layer of insulation can be treated in the same way, but the efficiency of such insulation is not consistent over the whole area of the floor, due to the fact that the heat loss, likewise, is not consistent. Therefore the cost of overall insulation is unlikely to be justified.

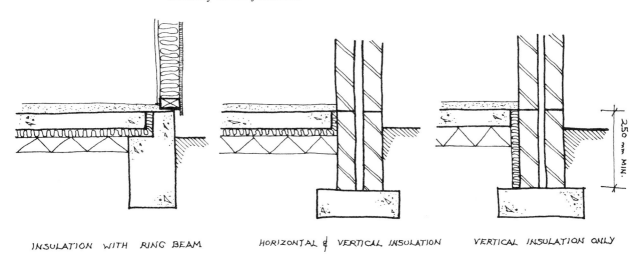

INSULATION WITH RING BEAM   HORIZONTAL & VERTICAL INSULATION   VERTICAL INSULATION ONLY

Figure 5.7   Insulation of ground floors next to earth

Perimeter insulation, however, is another matter – indeed, as we shall see later, it is often essential in order to avoid condensation on the floor surface. There are two ways of carrying out perimeter insulation (Figure 5.7). If vertical insulation alone is to be used it should be a minimum of 250 mm deep, and could with advantage be taken down much deeper, even to the top of the foundations. Horizontal insulation, on the other hand, should be about 1 m wide and should be combined with a vertical strip of insulation the full thickness of the floor slab around its exposed edges. Insulation in these positions must be of a type unaffected in performance or durability by moisture, such as closed cell polystyrene; alternatively, it should be protected from ground moisture by a damp-proof membrane.

Correction percentages to be used to adjust the values in Table 5.11, to compensate for the effects of vertical edge insulation, are given in Table 5.12. Horizontal insulation will be equally as effective as similar depths of vertical insulation.

Dimensions of floor (m)	Percentage reduction in U value for edge insulation extending to a depth of:		
	0.25 m	0.5 m	1.0 m
Very long× 30	3	7	11
× 15	3	8	13
× 7.5	4	9	15
60× 60	4	11	17
30× 30	4	12	18
15× 15	5	12	20
7.5× 7.5	6	15	25
3× 3	10	20	35

Table 5.12   Correction factors for perimeter insulation to be applied to the values in Table 5.11

Dimensions of floor (m)	U values for timber floors			
	Basic thermal resistance (any floor)	Bare or linoleum, plastics or rubber	With carpet or cork	Any surface with 25 mm quilt
	$R_{si}+R_a+R_e$ $(m^2\,°C/W)$	$(R_s = 0.20)$ $(W/m^2\,°C)$	$(R_s = 0.26)$ $(W/m^2\,°C)$	$(R_s = 0.86)$ $(W/m^2\,°C)$
Very long× 30	5.35	0.18	0.18	0.16
× 15	2.82	0.33	0.33	0.26
× 7.5	1.67	0.53	0.52	0.37
150× 60	7.13	0.14	0.14	0.12
× 30	4.58	0.21	0.21	0.18
60× 60	5.90	0.16	0.16	0.14
× 30	4.02	0.24	0.23	0.20
× 15	2.54	0.37	0.36	0.28
30× 30	3.42	0.28	0.27	0.22
× 15	2.34	0.39	0.38	0.30
× 7.5	1.55	0.57	0.55	0.39
15× 15	2.03	0.45	0.44	0.33
× 7.5	1.44	0.61	0.59	0.40
7.5× 7.5	1.27	0.68	0.65	0.43
3× 3	0.75	1.05	0.99	0.56

Table 5.13   U values for suspended timber floors immediately above ground and basic resistances applicable to any floor structure

Suspended ground floors   A floor suspended above an enclosed air space is exposed on both faces to air. The air below the floor, because the ventilation rates in a normally ventilated under-floor space are very low, is at a higher temperature than the outside air. Table 5.13 gives basic thermal resistances common to all suspended floors. These figures are made up of the sum of R_{si} (inside surface resistance), R_a (the resistance of the air space ventilated by 2000 mm² gaps per linear metre of boundary) and R_e (the resistance of the earth). To these must be added the resistance of the floor structure (R_s) and any added insulation. Three specifications of timber floor are given in the headings to the three right-hand columns of Table 5.13 with R$_s$ values of 0.20, 0.26 and 0.86 respectively, reading from left to right. The U values of these structures are given in the columns below. All other structures, or the ones specified with additional insulation, will have to be assessed either by a new calculation of the resistances of the structure, plus the common resistances of a suspended floor, or by additional resistances added to the reciprocal of the U value obtained from the right-hand columns, thereby establishing a new U value.

Additional insulation is often incorporated in timber floors, either as a continuous layer of semi-rigid or flexible material over the joists, or as semi-rigid material between the joists. In the case of concrete floors it is more usual to place the insulation on top of the structural floor (Figure 5.8).

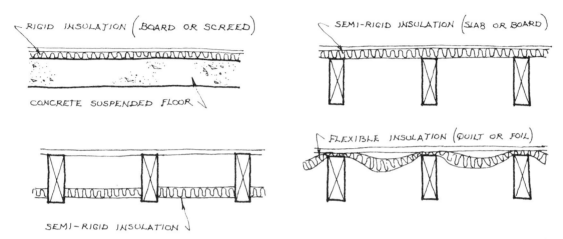

Figure 5.8   Insulation of suspended ground floors

It should be remembered that in any assessment of new or revised U values for suspended ground floors, allowance will need to be made for any new air spaces introduced.

Where blankets or quilts are laid over the joists they will be effective only over their uncompressed areas between the joists. Similarly, foils are effective only when used in conjunction with an air space; that is, between joists. In both these cases, though, the resistance of the joist itself compensates for these weaknesses.

Intermediate floors

The thermal transmittance values of intermediate floors can be calculated by the methods given above.

Windows

The U values for single and multiple vertical glazing and roof glazing are given in Table 5.10. These values are true for the glazing only. The effect of the frames will need to be calculated in order to assess the overall U value of the glazed area. The cold bridge produced by metal frames, which are made up of 'through' units exposed to both internal and external environments, should not be ignored when one is considering either heat loss, or condensation. Thermo-barrier units which avoid through members, though more expensive, are well worth the extra cost (Figure 5.9).

U values for typical windows with average areas of frame are given in Table 5.14.

THE EFFECT OF INSULATION ON INTERNAL SURFACE TEMPERATURE

In Chapter 2 it was explained that in order to provide a comfortable internal environment the surface temperature of internal spaces needs to be maintained no lower than 3°C below the air temperature. Indeed, the most pleasant conditions prevail when the surface temperature is slightly in excess of the air temperature.

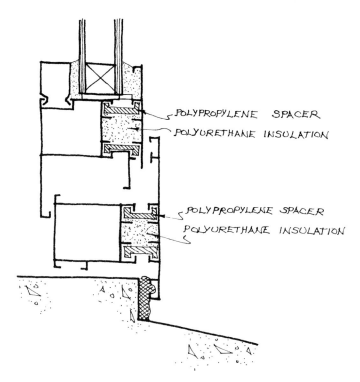

Figure 5.9 Typical thermo-barrier aluminium window frame

Window type	% of area occupied by frame	U values for stated exposure (W/m²°C)		
		Sheltered	Normal	Severe
Single glazing:				
Wood frame	30	3.8	4.3	5.0
Metal frame	20	5.0	5.6	6.7
Double glazing*				
Wood frame	30	2.3	2.5	2.7
Metal frame with				
thermal break	20	3.0	3.2	3.5

Where the proportion of frame differs appreciably from the above the U value should be calculated.

* 20 mm air space between glass.

Table 5.14 U values of typical windows

Insulation can obviously aid the achievement of this standard. It is advisable, therefore, to check the insulation values of the surrounding structure – insulated already, probably, for other reasons – in order to be assured that the overall thermal resistance of the envelope is sufficient to satisfy this condition of comfort.

Assuming a steady state of heat flow, the insulation requirements to satisfy this criterion can be calculated from the following expression:

$$R = \frac{R_{si} \times \theta d}{3°C}$$

in which R = the total thermal resistance of the structure (m²°C/W)

R_{si} = internal surface resistance (m²°C/W)

θd = temperature difference (°C)

The graph in Figure 5.10 shows the relationship between room temperature and levels of insulation based on this 3°C difference, when the design outdoor temperature is 0°C.

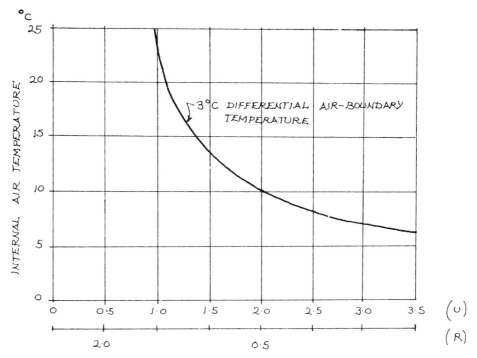

Figure 5.10   Graph of U and R values to give 3°C differential air-surface temperature

THE APPLICATION OF U VALUES

The earlier part of this chapter has mostly been concerned with the method of assessing the insulation characteristics of elements of structure. The comparative use of U values is clearly invaluable to the designer to assure himself that the 'overcoat' effect of his structure is going to be adequate. Indeed, mandatory insulation standards are based on the use of U values (see Appendix 1).

When steady state conditions apply (that is, when there are no marked fluctuations of internal or external temperature), the combined U values of the elements of the shell can be used to assess accurately the amount of heat needed to preserve the internal environment at a particular 'artificial' temperature. Two other factors need to be known as well:

1. The difference in temperature between the internal and external environments. For design purposes in the UK this is usually the difference in temperature between the internal design temperature of the space in question (Table 2.2) and some proposed very low external temperature. This then has the effect of ensuring that the heating plant will be of sufficient capacity to maintain the artificial temperature perpetually at the design temperature, in spite of excessively low external temperatures. The fact that the external temperature will fluctuate is ignored for the purposes of this calculation.
2. The loss of heat that will occur as a result of necessary ventilation. For preliminary design purposes this can be taken as an average over a 24 hour period (again a simplification of the actual conditions which will prevail) at the design ventilation rates appropriate to the internal space (see Table 2.2). This can be reasonably assessed from:

$$C_v = \frac{1}{3} N_v (t_c - t_{ao})$$

in which C_v = ventilation heat loss (W/°C)

$\quad\quad\quad$ N = number of air changes per hour

$\quad\quad\quad$ v = room volume (m³)

$\quad\quad\quad t_c - t_{ao}$ = the difference between internal comfort temperature and external
$\quad\quad\quad\quad\quad\quad\quad$ design temperature (taken as a low sol-air temperature)

Given these three pieces of information, the heat requirement can be established from the following equation:

$$\bar{q} = (\Sigma\ AU + C_v)\ (t_c - t_{oe})$$

in which $\bar{q}$ = the heat requirement in steady state conditions of heat flow (W)

$\quad\quad\quad \Sigma AU$ = the sum of the areas of the building shell multiplied by their
$\quad\quad\quad\quad\quad\quad\quad$ appropriate U values.

This equation holds good when temperatures do not change significantly with time. It is ideal for calculating heat flow during typical UK winter conditions, when there is little or no solar gain and long periods of low external temperatures. It does not, however, produce a realistic evaluation when temperatures, either internal or external, fluctuate. Therefore, when intermittent heating is applied to the internal space, or when external temperatures are influenced by direct solar radiation, another method of assessment needs to be used.

ADMITTANCE PROCEDURE

The shortcomings of the U value method, particularly applied to summer conditions, when internal temperatures or cooling loads need to be calculated, have led to the development by the BRE of a technique known as the admittance procedure. Not only does this cope with cyclic energy input, internal or external, but it also reflects the way in which a structure responds to the adjacent temperatures. For instance, in intermittently heated spaces the wall surfaces will more quickly respond to heat input if the layer closest to the heat source has a low thermal transmittance (ie, has good insulating properties). Alternatively, if the same layer is positioned furthest from the heat source, with layers of higher transmittance (brickwork or dense concrete) closest, the heavyweight material will *admit* the heat, conducting it through its thickness, and its surface temperature will consequently be slower to respond to heat input. Both walls could have precisely the same U value, their difference merely being that of the sequence of layers and not a difference of materials. Their air-to-air transmittance in steady state conditions would be identical, but their response to cyclic input would be totally different. It is this characteristic, too, that the admittance procedure recognises.

Admittance, like transmittance, is a characteristic of the building shell and expresses numerically its ability to store or release energy over the course of a period of time. It is, one may say, the alternating equivalent of thermal transmittance.

Swings of temperature about the mean are related to swings of energy about its mean. A very similar equation to the steady state heat loss calculation above expresses the instantaneous variation of energy input from its mean:

$$\tilde{q} = (\Sigma AY + C_v)\ \tilde{t}$$

in which $\tilde{q}$ = the cyclic energy input at a particular time (as opposed to $\bar{q}$, being the
$\quad\quad\quad\quad$ steady state energy input)

$\quad\quad\quad$ Y = the admittance of the surface (W/m²°C)

$\quad\quad\quad \tilde{t}$ = the instantaneous variation of temperature from its mean (°C).

It should be noted that $\tilde{q}$ and $\tilde{t}$ are true for a particular moment in the cyclic fluctuation of energy or temperature about the mean. Therefore, to establish fully the heat flow patterns, the calculation must be repeated for different times of the day.

Admittance values are given in Table 5.10. It will be noted that Y values lie in a similar range to U values (0 to 6 W/m²°C) and that their size depends on the surface of the component in question and its character of admitting or excluding heat. High admittance materials, such as thick masonry, lead to small temperature swings for a given energy input. Low values, typified by suspended ceilings, carpeting etc, lead to high temperature swings, or a quick reaction to energy input. It is the surface that is 'seen' by the heat source that determines the admittance value. It also, unlike steady state calculations, involves all room surfaces, not just the internal surfaces of external elements.

Two other factors are given in Table 5.10 – *surface factor* (F), which is the proportion of the heat gain at the surface, subjected to the cyclic energy input, which is readmitted to the adjacent space, and *decrement factor* (f) which is a ratio of the cyclic transmittance to the steady state U value – or, in other words, how much heat penetrates the structure. Obviously a time lag is also involved in all these heat transferences which can vary from nothing to two hours depending on the 'weight' of the material. All this has to be taken into consideration when calculating the cyclic heat transmittance – a very complicated procedure and not one the building designer will want to become involved in. It is a job for the professional. The building designer merely needs to know the language with which to discuss the problem with his services engineer.

THERMAL EVALUATION AT AN EARLY DESIGN STAGE

As we have seen, the sophisticated methods now evolved for calculating the thermal performance of the building shell can be very complex, not to say tedious. The calculation of hourly values of $\widetilde{q}$, involving fluctuation of energy input from the sun, the occupants of the building, its lighting and equipment, as well as internal heating and cooling cycles, is an operation more fitted to a computer than manual calculation.

The problem, however, is that most existing programs have been produced to satisfy the present needs of the services engineer. As we have seen, the services engineer (however undesirable this may be) is not used to being involved in the initial design stages of a building. Instead, he expects to be given the problem of heating or cooling a building whose design is already largely determined. He then applies the available procedures to evolve an economic solution, given the criteria with which he is presented. The available programs have been developed to satisfy this need. They often require more detailed data than are available early in the design process. They also require of the architect an expensive involvement with a computer bureau.

The obvious necessity is for a simple method of evaluation which the architect can apply early in his design stage, to assess the likely thermal performance of his proposals. Such a method has been evolved by the BRE, in collaboration with the DOE and the DES. At the time of writing it is scheduled to be published by the HMSO some time in 1979.

The method consists of a series of graphical aids; just one example is given here (Figure 5.11). In a building with a low ratio of surface to floor area (typical of an office, school, or hospital) thermal insulation and solar gain will mainly be decided by the type and size of the windows and their shading devices. By allocating values to energy input from the occupants and the lighting, and a value for thermal response, it is possible to establish relationships between inside temperature, ventilation and area of glazing for a given combination of glass and shading. These conditions are shown in the first graph in Figure 5.11, which serves to indicate desirable ventilation rates. The likelihood that these will be achieved by natural means (by either buoyancy or wind forces) can be assessed by the second and third graphs, which will be produced as overlays and can be used to establish proportions of windows that can be opened and the height of the window opening. A daylight predictor is given in the fourth graph overlay.

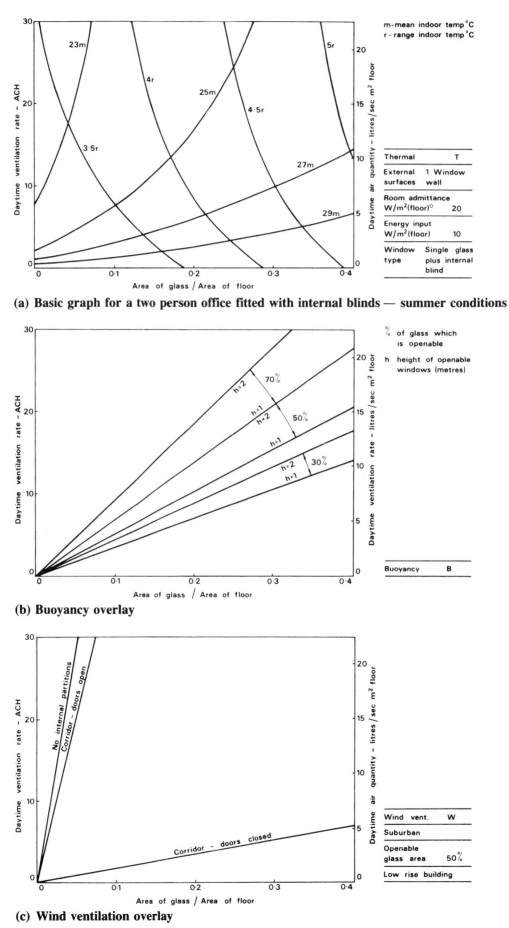

(a) **Basic graph for a two person office fitted with internal blinds — summer conditions**

(b) **Buoyancy overlay**

(c) **Wind ventilation overlay**

Figure 5.11   Example of BRE thermal design aid

These graphical aids reverse the computer process, which predicts thermal performance from a designed building, by defining the thermal performance and then checking the building design against it. The designer can see where his options lie and what benefit can be obtained by, maybe, relaxing a performance requirement occasionally. It also has the merit of speed and ease of application.

6   The causes of condensation

The problem of condensation is very much a problem of today. A few years ago it was hardly ever mentioned; now a week rarely goes by without headlines in the technical press concerning alleged abnormal condensation – usually in some local authority housing project. In fact, the problem has become very much associated with domestic buildings. In extreme cases it can cause considerable concern, expense and even discomfort to those who are unfortunate enough to experience it. Others never seem to encounter the phenomenon, except in a little hazing up of windows during cold, winter weather.

All building types can be prone to condensation, given the conditions necessary for its formation. Yet it is in housing, particularly, where the problem is reaching crisis proportions. Some specialised buildings also suffer; but always in a way which can be anticipated because of the activities taking place in the buildings. Yet we have been living in houses for hundreds of years without, apparently, too much discomfort from condensation. Therefore we might be excused for not having anticipated the problem in such a well-known (and maybe, therefore, too-little-respected) type of building.

Condensation results from a series of relatively simple, totally invariable and well understood physical factors. Its occurrence should, therefore, be thoroughly predictable. So why does it take us by surprise?

We shall see that today's problems of condensation have resulted from a lack of appreciation of the effects of changes in our living habits, as well as of some design aspects of present-day housing. For instance, many of the types of heating installed nowadays in low cost housing aggravate, rather than ameliorate, the problem.

Condensation is very much related to the way in which we heat, ventilate and insulate our building shells. Insulation has not, until recent years, figured large on our shopping list of essential characteristics of the domestic building shell. Following a phenomenal increase in the cost of fuel, this omission has been rectified. Heating itself used not to be an immensely expensive item in the family budget. Indeed many families allowed themselves very little house heating. Today, that situation would be unacceptable, due to the higher standards of thermal comfort expected by the majority of people, who like to have the whole house heated to high thermal standards – but only if such results can be achieved at minimal cost. This drives them to expedients which are often disastrous. The situation is further complicated in Britain by the government-sponsored 'save it' campaign, which in encouraging the saving of all types of fuel often discourages the allowance of very necessary fortuitous air infiltration to buildings.

In any consideration of the thermal performance of the building shell, it is essential to consider this related factor of condensation. Not only does the imposition of extra insulation substantially alter the risk of condensation occurring, it should be remembered that the effects of condensation can cause almost as much concern to the agile adult as do moderately cold conditions. It is not a minor inconvenience. In extreme circumstances it could cause structural collapse.

HOW CONDENSATION OCCURS

Water vapour, in varying quantities, is always present in the air. It is rarely visible and, for the majority of the time, we cannot even sense that it is present. We only become aware of humidity when cold, dry air meets warm, moisture-ladened air and fog results, or indirectly when inside our buildings a rather nasty black mould starts growing on the walls of our less well-heated rooms, or even on our clothing hanging in built-in cupboards on external walls.

We may also become conscious of a humid atmosphere on hot days when we find the heat oppressive out of proportion to the actual air temperature. This condition is indigenous to many parts of the world, which have warmer climates than the UK and experience consistently high humidities. The coastal regions of the Red Sea are particularly good examples of this. The high moisture content of the air results in its inability to absorb more moisture, and therefore one's body cannot lose heat through the evaporation of sweat, particularly if there is little air movement. The result is over-heating of the body and the sensation of a higher air temperature than is actually being experienced. In these conditions even newly washed clothes or other articles take a long time to dry, unless actually hung in the direct sunlight.

But these are abnormal conditions by the standards of the temperate parts of the world. With moisture contents between 40% and 70% we rarely notice the humidity. Ideally, internal conditions should be maintained within these extremes. Lower humidities cause dryness of the throat; higher humidities, feelings of oppression. Condensation (which is the visible manifestation of the water vapour in the atmosphere) is, however, a very common occurrence. It can be experienced in the UK whenever the external air temperature drops towards freezing point. Why is this?

Warm air is capable of carrying more water vapour than cold air. When warm air comes into contact with either colder air or a cold surface, the warm air is cooled. If in its original state it had a high humidity, it is likely that this cooling will depress the temperature of the air to a level at which it can no longer contain all the water vapour it was holding, and some of it will be discarded as condensation. This takes the form of thousands of minute particles of water, either suspended in the air, as in the case of fog, or as a dew on cold surfaces. On fine, clear days we can see the condensation of the airborne water vapour forming as dew after sunset on those surfaces which are radiating their heat towards the clear, night sky, causing a sharp drop in their surface temperature. The adjacent air is discarding some of its water vapour. An example indoors is that of a single glazed window on which, on cold nights, moisture will form. The glass, being a good conductor of heat, quickly acquires the external air temperature. The adjacent internal air is then cooled and some of its water content condenses out on the glass.

Water vapour in the air exerts a pressure, which contributes to the total pressure of the air. The more moisture present in the air, the greater the contribution of water vapour to the total pressure of the air. This contribution is known as vapour pressure and is measured in millibars (mb). To relate this unit to other metric units $1 \text{ mb} = 10^2 \text{N/m}^2 = 100 \text{ Pa}$.

Air inside a heated building usually contains more moisture than does the external air. Hence it is at a higher pressure and therefore the pressure tends to force the warm air through the structure, if it is permeable, taking the moisture with it.

The moisture content of air is expressed either in kg/kg of air, or g/kg. This is the ratio of the weight of water vapour present in the air to the unit weight of dry air (1 m³ of dry air weighs 1.20 kg).

Relative humidity

The relative humidity of air (RH) is expressed as a ratio between the actual vapour pressure of a sample of air at a particular temperature and the maximum vapour pressure the sample could contain at the same temperature. Alternatively, RH can be regarded as the amount of water vapour present in the air expressed as a percentage of the amount of water vapour that would be required to *saturate* the air at the same temperature. Therefore the RH depends on the amount of moisture in the air and the temperature of the air. With constant moisture content, the greater the temperature of the air, the lower the RH; conversely, if the RH remains constant, the higher the temperature, the greater the actual amount of water vapour held in the air.

As an example, if we take a series of external air temperatures, typical of minimum winter temperatures in the UK, and examine the moisture contents of the air at 90% RH, we arrive at the following figures:

External air temperature °C	Moisture content of air kg/kg	Vapour pressure mb
0	0.0034	5.3
−3	0.0027	4.3
−5	0.0022	3.6

Air is said to be *saturated* when its RH reaches 100%; in other words when it cannot contain any more water vapour at the existing temperature. Air at 14°C is saturated when it contains 10g/kg; air at 18°C when it contains 13 g/kg. If the temperature of the air falls until saturation point occurs, the air is said to be at its *dew point* and a further fall in temperature will result in water vapour being forced to condense out. The amount of water vapour condensing will be the equivalent of the amount of vapour in excess of 100% RH of the air at its new temperature.

Condensation in a building usually occurs when warm air comes into contact with a cold surface, which cools the air below its saturation point, causing its excess water vapour to change into liquid water. This can happen on a plastered wall, or a pane of glass. In fact, it can happen on almost any surface, including that of fabrics. Condensation is, however, more obvious on the harder, more impervious surfaces, such as glass, glazed tiles or gloss painted plaster. This is called *surface condensation*. But condensation can also happen within the thickness of a building component, due to the internal air permeating through the structure because of its greater pressure, and meeting a cold area within the structure. This is potentially the greater hazard as the resulting dampness can often go undetected until substantial damage has been caused. This type of condensation is known as *interstitial condensation*.

Most building materials, except metals, plastics sheeting and boards faced with metal foil, are to some extent permeable and therefore do not obstruct the diffusion of moisture-ladened air through the structure. This air will cool as it moves through the structure and may eventually cool to below its dew point. This is particularly likely to occur when the construction is faced externally with an impermeable layer. A typical example of this is a building whose walls are clad with metal sheeting and whose internal surface is a decorated insulation board. For this structure to satisfy minimum standards of thermal insulation it will probably contain within its thickness lightweight insulation (mineral wool or glass fibre, for instance). Neither the insulation board nor the additional insulation will form any barrier to the moisture-ladened air. As the warm air moves, therefore, through the insulation there will be a marked temperature drop and a risk of condensation forming on the back of the metal cladding. This surface has a high vapour resistance and, in effect, acts as a barrier to any further outward movement of the air, thereby also preventing any chance of the condensation drying outwards. As a result, there is a very real danger of the condensation dampening the insulation, with the result that its insulating properties will be largely impaired. The only way to alleviate this is to provide adequate ventilation behind the cladding to disperse the condensation and dry out the insulation. The effectiveness of this is, however, open to question, and therefore the best course of action would be to try to inhibit the penetration of the warm air from inside the building by the provision of a vapour check on the warm side of the insulation. The question of vapour checks will be examined in more detail later (see Chapter 7).

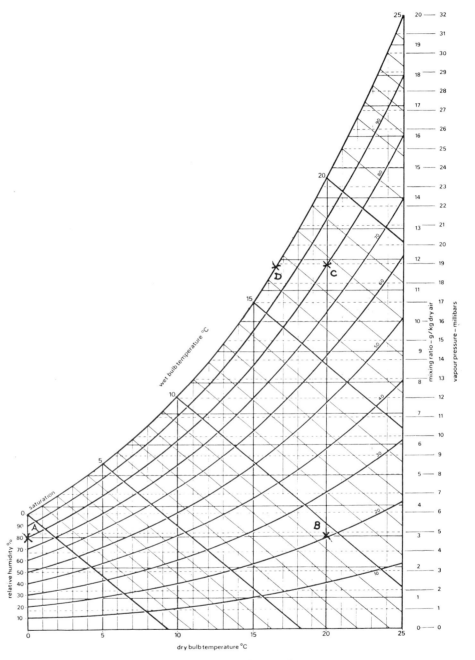

Figure 6.1   Graph relating relative humidity to air temperature

From this example it will be understood that interstitial condensation is dangerous, not only because it can cause structural failure if there is any element within the thickness of the structure, such as timber, which is vulnerable to rot or decay in damp conditions, but also because the moisture is likely to render ineffective any insulation inside the component. This can have disastrous effects on the carefully calculated thermal efficiency of the building shell. It is likely, also, further to aggravate the condensation problem.

Surface condensation is much more evident, due to discolouration of decorations and the formation of mould growths on paint, wallpaper and textile surfaces. Such mould growth is likely whenever the RH is frequently, and for prolonged periods, in excess of 80%. But at least surface condensation can be easily seen and the physical conditions causing it rectified.

To summarise, any surface or plane within a permeable element will be subject to condensation when its temperature is below the dew point temperature of the adjacent air.

Figure 6.1 shows a graph relating air temperature and moisture content to RH. If we apply the information on this graph to an example, we shall see how the two related conditions of temperature and moisture content produce situations in which condensation is unavoidable. External air at a temperature of 0°C and a RH of 80% is not an uncommon winter condition in the UK. This is point A on the graph. From the right hand axis it can be seen that the moisture content of the air is 3 g/kg. Due to natural ventilation processes this air enters a living room heated to 20°C. Eventually it too will gain this temperature. If, in the process, it has not altered its moisture content its RH will have dropped to just over 20% – point B on the graph. This is very unlikely. Because of the relatively humid internal atmosphere, caused by the breathing of the occupants and their other moisture-raising activities, such as washing and cooking, it is quite possible that the air could return to 80% RH; but now at its higher temperature this would mean a moisture content of 11.8 g/kg – point C on the graph. For this air, then, to be in a state of saturation it would be necessary for it to be cooled to 16.5°C or below (point D) and, therefore, should this air meet any surface at this or a lower temperature it would be caused to give up some of its water vapour as condensation. This serves to emphasise the importance of maintaining internal *surface* temperatures not far below the *air* temperature of the internal space (see Chapter 2).

The factors which effect the formation of condensation can be stated as these:

1. The moisture content and temperature of the air entering the building.
2. The moisture content and temperature of the air inside the building.
3. The rate of ventilation.
4. The room surface temperature.
5. The temperature and permeability of the structure.

These factors we shall now examine in more detail.

FACTORS IN CONDENSATION PRODUCTION

Air entering the building

In cold weather the outside air is usually so cold that the moisture content of the incoming air is of little significance; that is, even with a high RH it contains a very small quantity of water vapour. The example given above illustrates this well. It is, therefore, safe to say that this factor can be ignored in cold weather. In summer, however, the situation can be different.

When there is a sudden change in temperature, from cold to warm humid weather, there will be a short period when condensation can form on cold structural surfaces until they have had a chance to warm up. This is a temporary phase known as *warm weather condensation* and usually affects only high thermal capacity structures which are slow to heat up. Ventilation, in this case, merely aggravates the problem. The only remedy is to increase the heat input to hasten the warming up process. Normally, though, the combination of conditions required to produce this phenomenon is shortlived and the resulting condensation will certainly do no lasting harm to a building's structure, or to its decorations.

In hot and humid parts of the world similar conditions can occur and may not be so transitory. In these cases condensation will possibly form on surfaces cooled by the air conditioning – but not on the inside surfaces as the internal air will have been dehumidified by the air conditioning. The danger here is that warm, moist air will diffuse through the structure from outside and cause interstitial condensation as it approaches the inner layers of the structure. This can be avoided by the use of vapour checks towards the outside of the structure – the reverse of the normal position in temperate parts of the world.

The amount of rainfall in an area does not seem to markedly affect the risk of condensation. External humidity in temperate climates (with the exception of warm weather condensation) has little influence on internal condensation. It should, however, be remembered that condensation can occur when external air temperatures are considerably higher than 0°C. In many houses the risk is present with external temperatures of 5°C, or even 10°C. This is therefore not a problem that is restricted to an average of 30 days a year when, in the UK, temperatures can be expected to be below 0°C.

Air inside the building

If external humidity has little influence on condensation risk, internal humidity is clearly the significant factor. Considerable quantities of water vapour are emitted into the inside air in the average dwelling by the occupants and their activities.

For example, it has been estimated that four persons living in a house for 12 hours will, merely by the process of breathing, add 2.5 kg of water vapour to the internal air. An average person engaged in sedentary activities will exhale more than a litre of water vapour in 24 hours. More energetic activities can increase this amount by four times. Typical family activities – breathing, cooking, washing and drying clothes – produce about 12 litres of water vapour per day. Table 6.1 gives a list of some of the major causes of water vapour build-up in the internal atmosphere of a building. It will be noted that flueless heating appliances that depend on combustion to produce their heat (paraffin and gas heaters) are particularly prolific producers of water vapour. Natural gas produces 1.5 kg of moisture for each m³ burned; paraffin 1.3 kg for each kg of paraffin. Flueless heaters clearly increase the risk of condensation prodigiously in domestic buildings by this enormous input of water vapour. Similar risks are associated with flueless gas heaters used in some industrial buildings.

Certain industrial processes, such as brewing and dyeing, produce high levels of humidity; but here the risk of condensation is usually appreciated and specialist aids to combat the problem are employed. The same can be said of pool halls in swimming pool buildings which can regularly experience RHs in excess of 70%. Office and commercial buildings usually have lower humidities (about 40%) and school classrooms rarely more than 50%. By contrast the domestic kitchen with inadequate ventilation can experience, for short periods, a RH of 70%. It is for this reason that condensation has become thought of as a predominantly domestic problem.

In domestic buildings the rooms whose activities produce the most moisture are the kitchen and the bathroom. An effort needs to be made to restrict this moisture and prevent it from spreading around the remaining areas of the dwelling and, as far as possible, to exhaust the water vapour near its source.

It should be remembered that mould growth can occur if RHs remain consistently above 70%, even if there is no apparent liquid condensation. If the RH exceeds 80%–90% for long periods mould growth is certain to occur.

The heating of a building is important as warm air can contain more moisture than cold air. The water-ladened air in a warm building can then be removed by ventilation. Also a good standard of heat, maintained consistently, will warm up the internal surfaces and help to keep them above dew point. Public and commercial buildings tend to be heated to a higher standard and more consistently than domestic buildings. With their lower humidities they rarely experience condensation on an excessive scale. The domestic building tends to be heated intermittently, the greatest demand for heat being, usually, during the evenings when the family come home from work or school.

Generally, living spaces are heated to about 18°C during at least some part of the twenty-four hours; but the bedroom temperatures can vary widely. On an average they appear to be about 5°C lower than the living rooms. Some are hardly heated at all and can only be marginally above outside air temperature. It is this type of economy – very laudable from a fuel conservation point of view – that results in many of today's condensation problems.

Those parts of the dwelling which are not heated, or are inadequately heated, will be those which are most likely to suffer from condensation. Turning off the heat during long parts of the day, punctuated by high temperatures during short periods, tends to aggravate the problem. In fact condensation risk can be minimised more by heating a building consistently but at a low temperature, than by pumping into it short, sharp bursts of heat at a very much higher temperature.

Source	Moisture output per hour (kg)
1 person asleep	0.04
1 person (average domestic activity)	0.053
Electric cooking	0.70
Gas cooking	1.00
Gas heating (no flue)	0.13 per kW
Paraffin heating	0.09 per kW

Average dwelling (5 person house) moisture production per day

Source	(kg per day)
5 persons asleep, 8 hours	1.6
2 persons active, 16 hours	1.69
Cooking	2.80
Washing up, etc.	1.00
Washing clothes	4.50
Clothes drying (unvented)	5.00
Washing and bathing (average)	0.50
Total	17.09

Table 6.1   Sources of internal moisture

Quick reaction heating systems, such as warm air installations, make occupants feel comfortable quickly but are not effective at maintaining surface temperatures at a reasonable level, particularly if the structure is of high thermal capacity. Under these circumstances, condensation may well be unavoidable. If the structure has a low thermal capacity, quick intermittent heating is not likely to cause so much obvious trouble, but care needs to be taken to make sure interstitial condensation of dangerous proportions is not occurring where no one can see.

Rate of ventilation

As the outside air in temperate climates is likely to be at a lower moisture content than the inside air, it is theoretically possible to avoid all condensation by adequate ventilation. The problem is that although this used to be possible a few years ago, it is unlikely to be practical today.

Not very many years ago, doors and windows were generally less well-fitting than they are today. The majority of domestic buildings had suspended boarded ground floors with gaping cracks between the boards. Hardly any house had less than two flues or chimneys. Natural ventilation was thus induced, whether the owner liked it or not, to a level of up to 4 air changes per hour. Today, it is unusual for the average dwelling to have 2 air changes per hour.

Here again we are back to the paradox of present-day living. On the one hand fuel costs have risen out of all proportion to other living costs and yet, on the other hand, because of comparison with places of work and entertainment our norms of comfort have risen. Everyone knows that the greater the ventilation, the greater the heat necessary to replace that which is lost. As a result the average householder seals himself in his dwelling in the self-righteous belief that he is not only saving his own money, but doing the right thing by society as well. (All government propaganda is angled towards encouraging this very action). He therefore draughtproofs all doors and windows, and blocks up remaining flues or permanent ventilators. Couple this action with poor heating and the classic recipe for condensation is complete.

The other design consideration which must be remembered is that ventilation is effective only if it is consistent throughout the whole volume of the interior. Even with adequate overall ventilation, if stagnant air pockets are left behind, by-passed by the main air flow, local condensation can occur. If there is a danger of surface condensation, one of the first places the tell-tale mould growth will appear is behind furniture, where the air is relatively undisturbed.

Room surface temperature   If the temperatures of the surfaces of an internal space are allowed to fall below the dew point temperature of the adjoining air, surface condensation will result. We have already discussed the need for a room's surface temperatures to be maintained within a few degrees of the internal air temperature for reasons of comfort (see Chapter 2). This need is amplified when surface condensation is considered. For comfort we found that surface temperature should never fall more than 5°C below the air temperature. 3°C is probably a more acceptable level.

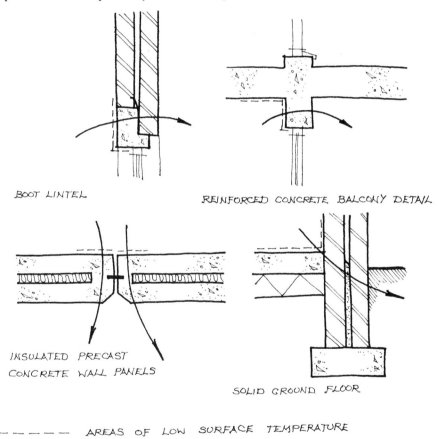

BOOT LINTEL

REINFORCED CONCRETE BALCONY DETAIL

INSULATED PRECAST CONCRETE WALL PANELS

SOLID GROUND FLOOR

— — — — —   AREAS OF LOW SURFACE TEMPERATURE

Figure 6.2   Some common examples of cold bridging

Assuming that the internal air temperature is reasonably high and that there is consistency of temperature throughout the building – that is, there are no unheated rooms to which the warm, moisture-ladened air can percolate – and assuming also that the shell of the building has reasonable levels of thermal insulation, surface condensation should never occur. Special attention, however, needs to be given to areas where cold-bridging is possible across the structure. This is an aspect which is often ignored. While the general U values of the enclosing elements are often adequate, designers have been known to forget about those areas in which different conditions apply, such as at lintels over openings, walls and ceilings adjacent to balcony overhangs, edges of ground floors next to earth, etc. (Figure 6.2). The lower temperatures of these surfaces can lead to local condensation.

Clearly, adequate insulation can minimise the risk of surface condensation by maintaining the surface temperature at a reasonable level; but it should be remembered that no amount of thermal insulation can make an unheated room warm. Insulation must always be complemented with a reasonable heating system if its effectiveness is to be realised.

If intermittent heating is to be used, the insulation should be close to the internal surface to achieve a quick surface reaction to changed internal temperature. If the building is heated more or less continuously (albeit to different levels over the twenty-four hour period) it does not matter where the insulation is placed within the thickness of the enclosing elements so long as its position does not encourage interstitial condensation.

BS CP 3: Chapter 11: 1970 gives a graph (Figure 6.3) showing the likelihood of surface condensation forming on internal surfaces, when the external temperature is 0°C and with varying U values for the enclosure.

The temperature and permeability of the structure

As we have discovered in the previous chapter, heat passes through an enclosing element and encounters three types of resistance on its way – the internal surface resistance, the resistance of the materials of structure and any cavities contained within the structure, and the external surface resistance. There is, therefore, a temperature drop across the structure. The position of insulation will affect the inclination of the temperature gradient across the element (Figure 5.1).

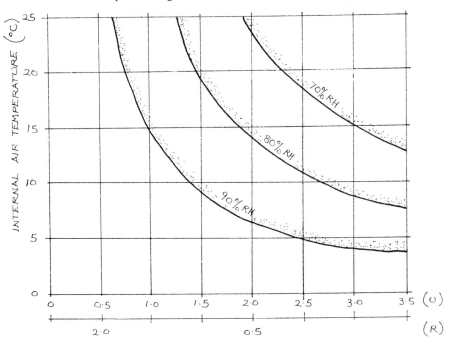

Figure 6.3   Insulation required to maintain the internal surface of the structure above dew point in steady state conditions with an outside temperature of 0°C

Assuming that the structure is permeable to water vapour (and most building materials are), moisture will be carried through the wall because of the water content in the internal air giving a pressure differential between the inside and outside air. The passage of this water vapour will also be subjected to resistance during its progress through the element, depending on the resistance of the materials of construction. If a complete vapour barrier could be installed at or near the inside surface of the enclosing element, it would be theoretically possible to eliminate the passage of the vapour within the structure. Under these circumstances, only surface condensation would be a potential problem, and if the surface temperature were maintained close to an adequately warm air temperature, the risk of condensation could be entirely removed.

In practice, as we shall see later, it is very difficult to provide a complete vapour barrier in most locations. As a result, when vapour passes through the element there is a reduction in vapour pressure, depending on the various resistances of the materials making up the element. If at any time, due to the falling temperature gradient through the thickness of the element, dew point is reached, interstitial condensation will occur (Figure 6.4).

This condensation need not necessarily be harmful if the materials of construction are not subject to rot, and if the overall thermal insulation of the element is not impaired. If, however, the element is constructed of a material that will rot or corrode in damp situations, or if the condensation is going to reach lightweight insulants, whose thermal performance will be decreased if they become damp, the viability of the whole structure is set at risk.

WHY CONDENSATION IS AN INCREASING PROBLEM

The reasons for condensation becoming an apparently increasing problem can be itemised in this way:

1. The way we live in our buildings.
2. Economic pressures.
3. Changes in building design.

The way we live in our buildings

It is widely accepted that the greatest area of concern about the growing problem of condensation is that related to domestic buildings. Admittedly there are problems in other buildings, but usually these are associated with some internal climatic abnormality which is generally recognised as causing high humidities – swimming baths, saunas and industrial buildings housing steamy, humid processes. These special buildings can be readily identified as having potential condensation

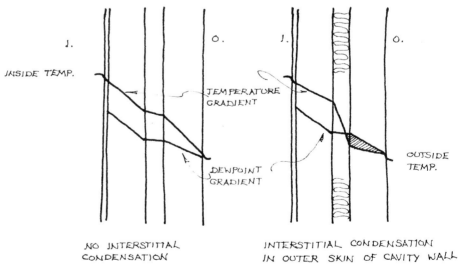

Figure 6.4 Temperature and dewpoint gradients through walls

difficulties and compensating action can be taken in the building design to ameliorate the conditions. The dwelling is another problem. Often the cause of substantial and persistent condensation is not immediately obvious.

For a start, the level of household moisture-making activity is often not appreciated. When this activity gets compressed into a very small part of the week, as happens in households where there is no one at home during the day for five out of every seven days, the output of water vapour reaches critical proportions during that very small time. Many dwellings are occupied only about 50% of every 24 hours – and seven of those hours the occupants of the dwelling are likely to spend in bed. In the remaining five or six hours most of the activities of the average household, which fifty years ago would have been spread over eight daylight hours, are compressed – washing, cooking, cleaning, bathing etc. Most of these activities are highly productive of water vapour. The fact that these activities are taking place during a substantial part of the year when it is cold and dark outside does not encourage the opening of windows to give the traditional 'airing' of the house that used to happen a few years ago.

The situation is further aggravated by changes in habits of personal hygiene and cleanliness. As recently as 1951 only 1.8 baths per week per person were taken in London and about 1.4 in the rest of the country[1]. This was undoubtedly due to there being only 56% of the dwellings in the UK with piped hot water a few years before, and only 46% with separate bathrooms. By 1970, however, fashions had changed to the extent that in some areas, such as the wealthy commuter belt of Surrey and Hampshire, two bathrooms were beginning to be thought of as the normal requirement for a three bedroom house. Improvement grants and the new stock of housing built during the preceding twenty years had dramatically changed the whole concept of personal hygiene. Hair washing had become a once-a-week or, in the case of women, often a twice-a-week habit. Such labour-saving devices of the household as automatic washing machines had led to washing becoming less of the drudgery it once had been and therefore becoming an activity indulged in more frequently.

Washing has become in many households an evening activity carried out in the kitchen, and it is unlikely that on cold winter nights, when there is the greatest risk of condensation, the kitchen will be properly ventilated. Also, as outside drying is probably going to prove difficult, due to the laundry having to be left unsupervised on the washing line overnight and perhaps all day, there is an increasing tendency for this activity to be almost completely performed indoors – often on heated airers or in tumble driers, without external air ventilation, in the kitchen. Alternatively the bathroom may be used for this purpose or, worst of all, sometimes an unused bedroom. The drying of laundry has always been a problem in high-rise flats where the facilities are usually inconvenient, inadequate or insufficiently secure.

Intermittent occupation also has the effect of encouraging the intermittent use of heating appliances. In addition, this type of occupation is going to encourage the use of quick reaction methods of heating, such as warm air systems, which produce comfortable air temperatures speedily, but whose operation is rarely long enough to heat up the structure of the building – particularly if the structure is of a high thermal capacity. Lightweight linings would react more quickly to the change in air temperature and help to eliminate this risk, but many traditionally built dwellings do not have such linings. The vast majority of families in this country are living in dwellings constructed of heavyweight materials.

In Chapter 2 the rising standards of thermal comfort were considered. Whereas in the early and mid 1950s the majority of people in the UK did not expect their houses to be generally heated, and tended to carry out their activities around one or two open fireplaces, today some form of general heating has become the norm. This has two effects. Firstly, because whole-house heating is expensive to run – today more than ever before – any heating system in a dwelling occupied by a lower income family is likely to be under-used and to be augmented by the use of cheaper-to-run paraffin heaters – and we have seen how much moisture they can produce. Secondly, the open fire has been almost entirely eliminated from current building designs. A few years ago the majority of houses with a piped hot water system could heat the water by means of a back boiler from a solid fuel appliance – ensuring that at least one flue was being used, thereby inducing ventilation to the house as a whole. Today the use of the back boiler is becoming rare. The old-fashioned chimney is disappearing, and as a means of introducing a considerable amount of fortuitous ventilation it is not being replaced.

Designers often excuse themselves from responsibility for condensation in their buildings on the grounds that it is the lack of intelligent usage of the building that causes the problem, not the design of the building itself. What they and we need to understand is that buildings should be designed for people – not people for buildings. If people wish to live in this way, some means must be found in the building design to accommodate this.

Economic pressures

The chief economic pressure – the cost of fuel to run a whole-house warming system – has been mentioned in the previous section. As fuel costs continue to rise, so the likelihood increases that heating systems will be less adequately used. In the case of old people or persons on fixed incomes, this is particularly likely. We have seen that the most effective way of avoiding condensation is to heat the dwelling consistently, rather than to expose it to short, sharp bursts of heat, but economic pressure tends to encourage intermittent use of heating systems. The attitude is, 'It's an unused room, so why pay to heat it?' or 'We're not at home all day, so we only turn the heat on in the evening'. Nevertheless the requirements to avoid condensation are to maintain internal surface temperatures and to avoid having some areas of the building at considerably lower temperatures than other areas. Otherwise, warm, moisture-ladened air percolates from the warmer parts of the dwelling to the colder parts where it deposits its moisture on the cold surfaces.

Another side-effect of economic pressure, as we have noted already, is that windows and doors are rendered more draughtproof by the occupants. Since the oil crisis in the early 1970s there has been considerable government pressure to draughtproof dwellings and thus avoid wasting fuel. They have promoted the idea on the basis of not only saving the country vast sums of money which would otherwise be spent overseas on oil, but also saving sums of money that will accrue to the householder. It is hardly surprising that the public have responded. Yet what they are in effect doing is, often, cutting off the natural ventilation routes which we have grown to expect over the years. This is something that needs to be very seriously considered. While it is eminently laudable not to waste fuel, one has got to ask whether the result of the 'seal-it-up' campaign is justified by the very small savings it makes. Modest ventilation does not cause great expense either to the householder or to the government. Its exclusion can lead to excessive condensation with all the structural risks involved. Is the small saving worthwhile?

Economic pressure also results in many householders being persuaded that the installation of double glazing will save them money. Given the usual window-to-wall ratio of the average house, it is unlikely that such action will save very much on the annual heating bill. It will, however, considerably reduce the condensation previously experienced on single glazing. Such condensation does, however, help to dehumidify the internal atmosphere. If the condensation that accumulates on single glazing is allowed to drain harmlessly away to the outside, it is arguable that single glazed windows perform a useful function, since condensation prevented from forming in one area is likely to appear in another.

Changes in building design

During the last thirty years changes have taken place in the way dwellings are designed. These changes particularly affect heating and ventilation.

The so-called Parker Morris minimal heating standards have been applied to all UK local authority housing schemes for a number of years. These require heating only to the living spaces (65°F, 18.3°C) and the kitchen and circulation areas (55°F, 13°C). Although these recommendations were originally conceived as minimal standards, rising building costs and restricted public spending have led to this being treated as a generally accepted heating norm for all but some old people's accommodation. This automatically produces dwellings of uneven heating and consequently creates a high risk of the unheated parts of the house – the bedrooms – suffering from condensation. This risk is often aggravated by bedroom doors being left open to allow the warm air from the rest of the house to circulate, bringing with it higher levels of humidity.

The degree of risk is increased in single level dwellings or in maisonettes with bedrooms below the living areas. These rooms are denied the warming influence of the heated living rooms, such they would receive in the traditional two storey house.

Design changes which affect ventilation include the omission of the open fireplace and chimney, already referred to, and the almost exclusive use of large-paned windows without easily controllable small-sized ventilators. Night ventilators in windows are much more likely to be used than are large pane opening lights, particularly large panes of the awning, top-hung, or pivot type. Not only are these large opening lights an all-or-nothing form of ventilation, but also they often lack adequate means of locking them in the open position. Thus they are a source of danger to children and provide a possible route for unauthorised entry to a building.

Permanent ventilators in rooms without a flue used to be demanded by most of the old local authority by-laws, but were omitted from the Building Regulations 1965 for England and Wales, probably reflecting an appreciation, on the part of those who drafted the regulations, of the fact that all permanent ventilators were usually blocked up by the occupier. The Scottish Building Standards, however, do still require permanent ventilation to habitable rooms, and maybe this is a desirable requirement. Some way needs to be found of providing this ventilation so that it neither inconveniences the occupier, nor lends itself to easy obstruction. Some forms of window ventilator satisfy these requirements.

THE CALCULATION OF CONDENSATION RISK

Irrespective of all the vagaries of design fashion or living habits, there are several invariable physical rules which govern the formation of condensation. When these are related to the proposed structure it is possible to calculate the risk of condensation. So long as a realistic view is taken of the humidity, temperature and ventilation rates applying inside the building, the principles can be used to assess the performance of all structures no matter what the occupancy pattern.

The risk of condensation occurring in or on enclosing elements is very much a matter of the thermal and vapour resistance of those elements, combined with internal humidity and temperature. When these facts are known, it is possible to calculate when and where condensation is likely to form.

First some basic design criteria need to be established.

Excess moisture content

The amount by which the internal moisture content exceeds that of the external air has to be assessed. Clearly this will vary, depending on the type of building and the sort of activities that are likely to take place in it. Shops, offices, classrooms, public meeting places and dry-process industrial premises – that is, buildings acquiring their excess moisture content from the bodies of their occupants – are likely to have an excess moisture content of about 1.7 g/kg; domestic buildings, because of the moisture created by cooking, washing and bathing and the possibility of restricted ventilation in cold weather, could be in the region of 3.4 g/kg; while catering buildings, or those housing humid industrial processes, could be as much as 6.8 g/kg. The differences in these figures indicate the reason for condensation occurring predominantly in domestic buildings (or in specialised, high-risk buildings) rather than in the average non-domestic building.

Vapour resistivity

Vapour resistivity, like thermal resistivity, is a property of the materials of construction and is the resistance offered by such material to the flow of vapour through unit thickness, measured in MNs/gm.

Vapour *resistance* relates to the resistance of a specific thickness of material and is measured in MNs/g. Usually the vapour resistance is quoted for thin membranes, such as gloss paint, polythene sheet or aluminium foil, while the vapour resistivity is quoted for other materials and their resistances are established merely by multiplying the resistivity by the material's thickness (in precisely the same way as thermal resistance is established from thermal resistivity). Table 6.2 gives a list of vapour resitivities and resistances for common building materials. Some authorities quote vapour permeability or diffusivity. These are the equivalent of the reciprocal of the resistivity.

Because condensation can occur either on the surface or inside an enclosing structure, these two risks need to be calculated separately.

Surface condensation   It has been established that the internal surface of an enclosing element has a surface resistance to heat flow (Chapter 2). Unless the structure is independently heated, the internal surface will be at a lower temperature than the internal air and consequently the air adjacent to the surface will be cooled. The amount of such cooling will depend on the thermal efficiency of the enclosing element – in other words, its resistance to thermal transmittance. The likelihood of surface condensation will depend on the relationship between the total thermal resistance of

Material	Vapour resistivity (MNs/gm)
Air – still	5.5
– with convection currents	negligible
Asbestos cement	1.6–3.5
Brickwork	25–150
Cement rendering	100
Concrete	30–100
Fibre insulation board	15–60
Hardboard	450–1000
Mineral wool	5
Plaster	60
Plasterboard	45–60
Plastics, cellular:	
expanded polystyrene	100–600
polyurethane (open or closed)	30–1000
urea formaldehyde	20–30
Timber – dry	45–75
plywood	1500–6000
Stone	150–450
Strawslab, compressed	45–75
Wood wool	15–40

Membrane	Vapour resistances (r_v) (MNs/g)
Aluminium foil	4000
Bitumen impregnated paper	11
Kraft paper, single	0.2
5 ply	0.6
Tar infused sheathing paper	0.6–190
Polyethylene, 60μm	110–120
100μm	250–350
150μm	450
Polythene 60μm	110–120
Paint:	
Emulsion – 2 coats	0.2–0.6
Flat oil – 2 coats	6–11
Gloss	7.5–40
Roofing felt	4.5–100
Vinyl-faced wall paper	5–10

Table 6.2   Vapour resistivities and resistances

the element and the surface resistance – or the temperature drop at the internal surface of the element.

In Example 1 given in Chapter 5 (page 65) the total thermal resistance of the structure was calculated as being 1.096 m²°C/W. We shall now examine this construction's performance with regard to the formation of surface condensation when the external air temperature is 0°C and its RH 100%. From Figure 6.1 we can read on the right hand axis that the external air would contain 3.8 g/kg of moisture, that is 6 mb vapour pressure. If we assume an internal temperature of 20°C with an excess moisture content of 3.4 g/kg – a normal excess for a domestic building – that proportion of the temperature drop across the wall element which is due to the internal surface resistance can be calculated from the following expression:

$$T_s = \frac{R_{si} \times T}{R}$$

in which T_s = temperature drop at the internal surface (°C)
R_{si} = internal surface resistance (see Table 5.3)
T = total temperature difference, inside to outside (°C)
R = total resistance of the element (m²°C/W)

Applying this formula to the construction in Example 1:

$$T_s = \frac{0.123 \times (20 - 0)}{1.096} = 2.24°C$$

The internal surface temperature is 20 − 2.24 = 17.76°C
The total moisture content of the internal air will be the sum of the external air moisture content plus the excess moisture content, or 3.8 + 3.4 = 7.2 g/kg.

Because the air immediately adjacent to the inside surface will assume the temperature of that surface, air with a moisture content of 7.2 g/kg will be cooled to 17.76°C. From the graph in Figure 6.1 we can see that air at this temperature can contain about 12.75 g/kg of moisture – considerably in excess of the actual vapour in the air. Therefore surface condensation will not occur.

If, however, the room being considered were an unheated bedroom at, say, 10°C, the situation would be very different.

$$T_s = \frac{0.123 \times (10 - 0)}{1.096} = 1.12°C$$

This would mean that the internal surface temperature was 10 − 1.12 = 8.88°C Therefore, the air adjacent to the external wall would be cooled to 8.88°C and from the graph (Figure 6.1) we can see that the air at this temperature can contain, approximately, only 7 g/kg of moisture. This is, therefore, a border-line case and condensation in some conditions is extremely likely. This example serves to emphasise the importance of maintaining general heating levels.

Interstitial condensation   The calculation of the risk of interstitial condensation is a rather more complex business. In this case it is necessary to establish the temperature drop through the thickness of the enclosing element and the possibility that moisture-ladened air, passing through the element, will reach a layer in the element whose temperature is below the dew point temperature of the air.

Taking once more the wall construction used in Example 1 (but this time annotating the planes of the element) the temperature and moisture content of the air at each one of these planes has to be calculated (Figure 6.5). The same internal and external temperatures and humidities will be assumed. First the thermal resistances through the element have to be calculated.

Thermal resistance through the element

Inside surface resistance, point 1, R_{si} = 0.123

Between points 1 and 2; 13 mm lightweight plaster

$$\frac{1}{0.19} \times 0.013 \qquad\qquad = 0.068$$

Between points 2 and 3; 100 mm concrete block

$$\frac{1}{0.19} \times 0.1 \qquad\qquad = 0.526$$

Between points 3 and 4; 50 mm cavity $\qquad\qquad = 0.180$

Between points 4 and 5; 105 mm brick

$$\frac{1}{0.73} \times 0.105 \qquad\qquad = 0.144$$

Outside surface resistance, point 5, R_{se} $\qquad = \underline{0.055}$
Total resistance (R_t) $\qquad\qquad\qquad\quad \overline{1.096 \text{ m}^2\text{°C/W}}$

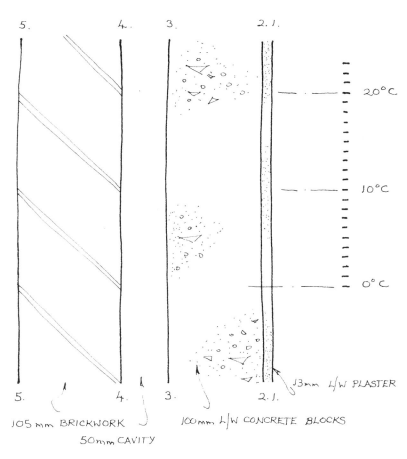

Figure 6.5   Example 1 wall construction with planes annotated

Temperature gradient   The total temperature drop from inside to outside (ΔT) is 20°C. The temperature drop at each plane of the element has to be established, using the formula:

Temperature drop (point to point) $= \dfrac{\Delta T}{R_t} \times r_t$

in which r_t = the thermal resistance of individual planes or components of the element already annotated above.
In the example in question:
Temperature drop (point to point) $= \dfrac{20}{1.096} \times r_t = 18.25\ r_t$ (°C point to point)

The temperature drop across the element can now be calculated.

Temperature drop across the element

Inside air temperature		=	20.0°C
Point 1;	18.25 × 0.123	=	2.24
			17.76°C
Points 1 to 2;	18.25 × 0.068	=	1.24
			16.52°C
Points 2 to 3;	18.25 × 0.526	=	9.60
			6.92°C
Points 3 to 4;	18.25 × 0.180	=	3.28
			3.64°C
Points 4 to 5;	18.25 × 0.144	=	2.63
			1.01°C
Point 5;	18.25 × 0.055	=	1.00
			0.01°C *

Outside air temperature = 0°C
  * error due to approximation.

The profile of this temperature drop should now be plotted against a temperature scale on the cross section of the external element (Figure 6.6).

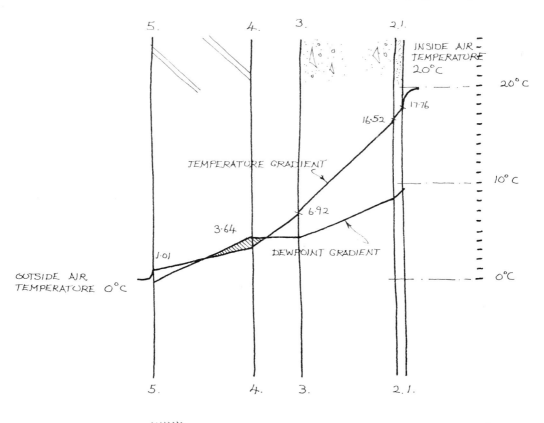

Figure 6.6   Thermal gradients imposed on Example 1 wall construction

Pressure gradient through the element   Referring to Figure 6.1 and using the same internal moisture content data given above, we find that the internal vapour pressure is 11.4 mb.

The vapour pressure difference (ΔP) is therefore 11.4 − 6.0 = 5.4 mb.

Using values for vapour resistance (Table 6.2), the vapour resistance at the various planes through the element can be calculated.

Vapour resistance

Points 1 to 2; 60 × 0.013   = 0.78
Points 2 to 3; 35 × 0.1   = 3.50
Points 3 to 4; nil   = 0
Points 4 to 5; 30 × 0.105   = 3.15
  Total vapour resistance (R_v) = 7.43 MNs/g

The pressure drop at each plane of the element can now be calculated using the formula:

Pressure drop (point to point) = $\dfrac{\Delta P}{R_v} \times r_v = \dfrac{5.4}{7.43} \times r_v = 0.73\ r_v$

in which r_v = the vapour resistance of individual layer of the element (MNs/g)

The vapour pressure drop across the element can now be established and this set against the corresponding dew point temperature of air with each reduced moisture content.

Vapour pressure drop across the element		*Corresponding dew point temperatures* (Figure 6.1)
Inside vapour pressure	= 11.4 mb	9.4°C
Points 1 to 2; 0.73 × 0.78 =	0.57	
	10.83 mb	8.5°C
Points 2 to 3; 0.73 × 3.50 =	2.55	
	8.28 mb	4.5°C
Points 3 to 4; nil =	0	
	8.28 mb	4.5°C
Points 4 to 5; 0.73 × 3.15 =	2.30	
	5.98 mb	0°C

The dew point gradient can now be plotted on the section which already contains the temperature gradient. Wherever the dew point gradient is above the temperature gradient, as it is in this case on the outer edge of the cavity and on the inside of the outer skin of the cavity wall, condensation will form. In the present example such dampness will do no harm. The outer skin of the cavity wall is expected to be wet, at some times in its life at any rate, and therefore the U value of the wall has been calculated on the basis of a damp outer leaf of brickwork. The condensation, therefore, will not degrade the thermal performance of the wall.

If we now look at the effect of adding insulation into the cavity (Example 3, Chapter 5) we find that the revised R_t would become 2.106 m²°C/W, giving an r_t value of 9.50.

The revised temperature drop across the element may now be calculated.

The temperature drop across the element with cavity insulation

Inside air temperature		= 20 °C	
Point 1;	9.50 × 0.123 =	1.17	
		18.83°C	
Points 1 to 2;	9.50 × 0.068 =	0.65	
		18.18°C	
Points 2 to 3;	9.50 × 0.526 =	4.99	
		13.19°C	
Points 3 to 4;	9.50 × 1.19 =	11.30	
		1.89°C	
Points 4 to 5;	9.50 × 0.144 =	1.36	
		0.53°C	
Point 5;	9.50 × 0.055 =	0.52	
		0.01°C *	

  * error due to approximation.

The vapour resistances at the various planes are for practical purposes unaltered, mineral wool having little vapour resistance. Had other forms of insulation been proposed there would possibly have been a greater total resistance. Only the temperature gradient on Figure 6.6 needs revising (see Figure 6.7).

Again, condensation is likely to occur in the same part of the wall thickness; but this time the area of risk is increased. This serves to illustrate that the addition of insulation can, on occasions, increase the risk of interstitial condensation. If the

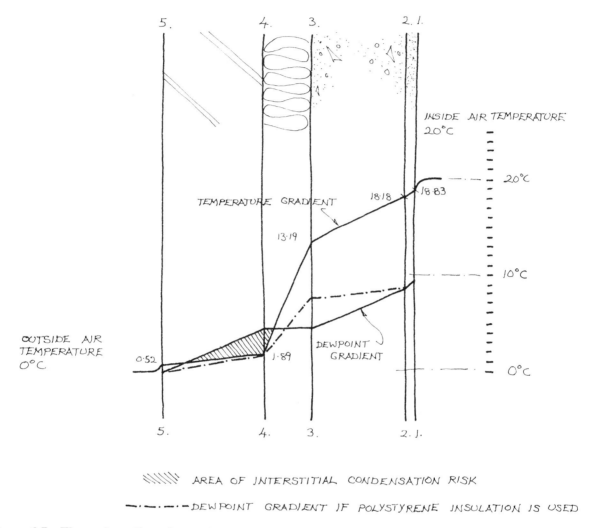

AREA OF INTERSTITIAL CONDENSATION RISK

— · — · — · —DEWPOINT GRADIENT IF POLYSTYRENE INSULATION IS USED

Figure 6.7   Thermal gradients imposed on Example 3 wall construction

insulation were of a type that would neither suffer harm, nor be degraded in its thermal performance if it became wet, there would be little harm. Mineral wool will, however, suffer in its performance and therefore the interstitial condensation has become undesirable.

If the dew point gradient could be altered by modifying the vapour pressure resistance, interstitial condensation would be avoided. The use of a different form of insulation with a vapour resistance, such as polystyrene, would serve to depress the dew point gradient below the temperature gradient – see Figure 6.7.

These two examples serve to demonstrate the principles which are involved in the assessment by calculation of condensation risk. It is advisable to check in all cases whether interstitial condensation is likely to occur. Often the designer concerns himself only with the U value of the enclosing elements and forgets completely about other related factors, such as condensation. Adequate U values prevent surface condensation; they need not necessarily do the same for interstitial condensation.

EXTERNAL DESIGN TEMPERATURES

It is important to establish the correct design temperatures for the external atmosphere for use in condensation calculations. BS 5250: 1975 advises that, while an external temperature of 0°C is a reasonable design temperature for most wall constructions, because of the substantial radiant heat loss experienced by metal sheet clad walls and lightweight sheeted roofs, it is safer with such construction to use lower design temperatures.

On clear winter nights, metal and asbestos-cement clad roofs experience a dramatic drop in temperature due to the degree of heat which they radiate towards the night sky. Often these structures can have an external temperature as much as 5°C *below* the outdoor temperature and this condition can be maintained for several hours. Clearly this imposes more severe demands on the thermal performance of the enclosing structure and this should be reflected in more stringent design temperatures. BS 5250: 1975 suggests −5°C should be used in the case of roofs and −3°C for sheeted walls. It may well be prudent to use the lower design temperature for all roofs with high emissivity roofing materials.

REFERENCES

1. Mather and Crowther, *Survey for the Coal Utilisation Council*, 1951.
2. Department of the Environment, *Homes for today and tomorrow*, HMSO, 1961.

7   The avoidance of condensation

If condensation is to be avoided in new buildings the designer must be aware of the ways in which his building design can influence the risk of condensation – in its planning, its heating and the design of its enclosing elements.

Firstly, the relationship of one internal space to another has to be considered with regard to the humidity expected in each and the pattern of air flow between them.

Secondly, the heating systems need to be designed so that they are capable of maintaining the temperatures necessary to discourage condensation without undue expense.

Finally, having regard to the expected humidity levels inside the building, the enclosing elements of the building shell need to be selected so that they are not likely to suffer either from surface or interstitial condensation.

Many condensation problems can be quite easily avoided, if the risk is appreciated at the beginning of the design process and the necessary action taken. Action attempted at a later stage may not prove as easy or as effective. Until recently the avoidance of condensation tended to be overlooked as one of the major requirements of the successful building. This omission is now rapidly being corrected, due to the increasing publicity that the problem has attracted in the last few years.

HIGH HUMIDITY AREAS AND AIR FLOW PATTERNS

The identification of the high humidity areas of a building is the first priority in the production of a condensation-free design. These areas should then be isolated, as far as possible, from the rest of the building.

Some of these high-risk areas will be instantly obvious – swimming pool halls, spaces housing humid industrial processes, abbattoirs, commercial kitchens, sterilising rooms in hospitals. Many will automatically be provided with mechanical extraction equipment. This has the effect of not only extracting the moist air, but also reducing the pressure in the risk area, thereby encouraging air flow from the less humid parts of the building and discouraging the dangerous flow in the reverse direction, which would spread the humidity around the building. Some of these high-risk areas will be provided with collection hoods over any steam-raising equipment – kitchen cooking ranges, sterilising sinks, industrial vats, etc. – with the object of controlling the spread of the hot and humid air and extracting it as near the source of the humidity as possible. This principle of dealing with the problem – or the cause of the problem – at source should be applied to all types of building including domestic buildings, whose high-humidity areas are the kitchen and the bathroom.

Having identified the areas of high humidity, these should be enclosed as far as possible from the rest of the building to discourage vapour circulation. Extraction should be arranged from the enclosure and the structural elements surrounding the humid room should contain vapour checks to discourage the permeation of water vapour through the structure to other parts of the building.

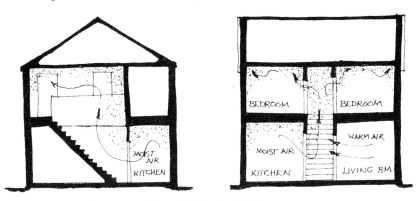

Figure 7.1   House layout and humid air circulation

Let us first examine how the application of these principles affects house design.

Domestic high-humidity spaces

The kitchen is the major risk area in the domestic building and should therefore never be planned as a part of an open living area, particularly if this is adjacent to the staircase. In fact the kitchen should always be planned as far away as possible from the staircase, although in the average small house this is an ideal that can rarely be achieved. The stair well is the chief means by which moisture-ladened air reaches the colder, bedroom area of the house. Providing a self-closing door to the kitchen would help to discourage too great a movement of humidity to the circulation areas. Figure 7.1 illustrates how house layout can influence the movement of water vapour about the dwelling.

While an easily openable window is an asset in a kitchen, if the window faces the prevailing wind the moist kitchen atmosphere can be blown back into the dwelling rather than being allowed to make its way outside. By far the most effective means of ventilating the kitchen is to provide an extract fan adjacent to the cooker and not too far away from the sink. An extract associated with a cooker hood is ideal. The extract fan not only reduces the humidity at source, it also reduces air pressure in the kitchen, encouraging a flow of air into the kitchen from other areas of the dwelling. This should be facilitated by the provision of a relief grille at low level from an adjoining space. The extract fan should always be placed at high level, as near the ceiling as possible and certainly above door head height. A 150 mm or 200 mm diameter fan will usually be sufficiently large for the average domestic kitchen and the cost of running such a fan is not excessive, either in terms of fuel or money. Its intermittent operation, when cooking or washing is taking place, would be roughly equivalent to the use of a 25 W light bulb. If a boiler requiring combustion air is situated in the kitchen, the provision of a relief grille is even more essential if the boiler is to function properly when the fan is turned on.

The enclosing surfaces of the kitchen should contain a vapour check to avoid water vapour passing into the structure. A gloss painted surface or a good vinyl paper would achieve this object. Where there is a special danger of interstitial condensation (in lightweight walls, or in ceilings immediately below a roof space) a vapour check in the form of a plastics membrane would be prudent.

Casual drying of washing indoors is a potential danger. This is becoming an increasing habit, particularly in flats which often have inadequate drying facilities. The problem can be largely overcome by the provision of a simple drying cupboard. This should be ventilated to the outside air at high level and have a low level inlet

107

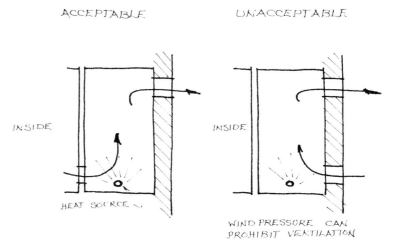

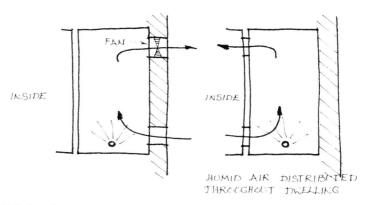

Figure 7.2 Drying cupboard

from the inside of the dwelling. It is unwise to design for the air inlet, as well as the outlet, to be external since this, under some weather conditions, can lead to ventilation not taking place, unless the outlet is fitted with a fan. Clearly the provision of a fan, whether the air inlet is internal or external, is always an advantage, but is not essential with an internal air inlet. (Figure 7.2). A self-closing door, again, would be an advantage and the internal finishes inside the cupboard should resist the passage of moisture into the structure.

Larders opening out of a kitchen can suffer from condensation. Good, permanent ventilation to the outside air is essential, both at high and low level, and the shelves should be set 15 mm to 20 mm away from the wall to allow air movement adjacent to the wall.

Another domestic room with potentially high humidity is the bathroom. Strangely, this room is not the hazard that the kitchen has proved to be. This is probably due to its more intermittent use and the fact that most people seem to appreciate the need to open the window to allow the steam to dissipate after a hot bath. The mandatory requirements for the ventilation of internal bathrooms are usually sufficient to avoid condensation, except when the room is used for long periods to dry clothes. Bathrooms on an external wall should be provided with a window with at least one small opening light, to allow fine adjustment of the ventilation and avoid draughts. This way the window is likely to be used. The provision of an electric fan would be an advantage, but is not usually necessary. A self-closing door should, however, be provided and the comments regarding vapour checks in the kitchen apply equally to the bathroom.

Bedrooms are often the rooms in which the effects of condensation are particularly marked. This is primarily the result of poor heating, as we have seen in the last chapter. However, ventilation can dramatically relieve the problem. The Building Standards (Scotland) Regulations require permanent ventilation to all habitable rooms (see Appendix 1) and this advice could well be followed generally, particularly with regard to bedrooms. The problem is that the majority of permanent ventilators are eventually blocked up by the occupants of the property. The ones least likely to suffer this fate are strip ventilators in windows (Figure 7.3). They should be sized at the rate of 300 mm² per 1 m³ of bedroom volume. In addition, it is

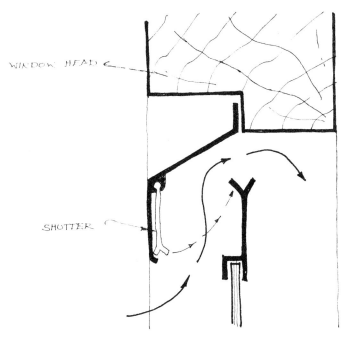

Figure 7.3   Strip ventilator in the head of a glazing pane

more likely that occupants will use their windows at night if they are provided with small night ventilator lights. This is both a psychological reaction and the result of a very real fear of insecurity.

There are other aspects of the design of dwellings which can affect the risk of condensation, but which have no immediate connection with the humid areas of the house.

The inability to induce cross-ventilation, as so often happens in flat layouts, can result in very low ventilation levels and pockets of stagnant air – always positions of condensation risk. Deep narrow rooms with a window on one short wall often suffer similarly from stagnant air pockets. Parts of buildings which have a greater exposure than others (flying bedrooms, top floor gable end flats, etc.) are always liable to greater condensation risk. Such areas should be provided with additional insulation in order to overcome their greater heat loss. Cold bridges, associated with concrete balconies and similar details, should be avoided. Internal walls between flats and permanently ventilated common circulation stairs with no heating should always be treated from an insulation point of view as external walls. (Figure 7.4).

Cupboards and wardrobes on external walls can suffer from condensation, often affecting the contents of the cupboard. Ideally all cupboards should be planned internally, but where this is not possible they should be well insulated to the external wall and ventilated at high and low level into the dwelling, possibly through the door.

Non-domestic buildings   Similar principles, involving the identification of the high humidity areas, their isolation and the control of the resultant humid atmosphere, apply equally to non-domestic buildings.

The examples are too numerous and too specialised to examine in detail here. Each case has to be assessed objectively, bearing in mind its particular problem. The solution will employ the same principles as those we have already discussed. We shall examine just one especially critical example – an indoor swimming pool.

The air temperature of the pool hall is usually maintained at about 28°C – similar to the temperature of the water in the pool, whose large surface area is continuously charging the air with considerable quantities of moisture through evaporation. The dressing rooms, however, are usually at a slightly lower temperature, probably around 24°C. There is still a considerable quantity of water in this area due to wet bodies, clothing and associated dampness. The entrance foyer, refreshment areas

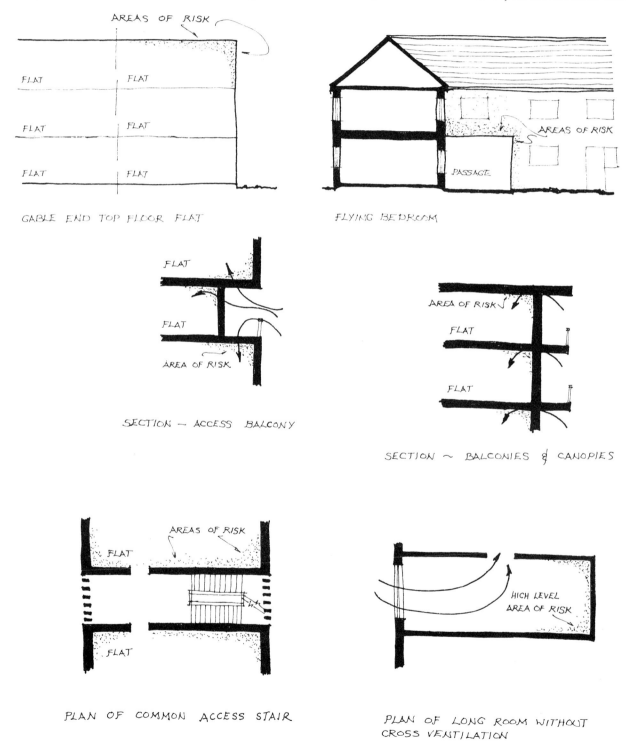

Figure 7.4   Areas of special condensation risk

and other public spaces are at more normal temperatures and humidities. It is obviously desirable to isolate the pool hall atmosphere from that of the other areas of the building. The same applies to a lesser extent to the dressing rooms. Infiltration from either of these areas to the public spaces would convey high humidities to less highly heated areas and would be, therefore, undesirable.

Swimming pool buildings today are usually designed with a ventilation system, or even a full air-conditioning system. As a result the solution is relatively simple. The system has to be designed to provide a slight negative air pressure in the pool hall, thereby encouraging air flow from the dressing rooms and public areas alike. Similarly the dressing rooms should be slightly below the air pressure of the public areas.

While the building is in use the operation of the ventilation plant usually avoids too excessive a build-up of humidity and thus prohibits condensation. However, when the plant is turned off during the night, evaporation continues from the surface of the pool and a build-up of humidity to dangerous proportions is possible. This is usually avoided by some extended running of the plant during hours when the pool is closed, but 24 hour running is unlikely to be necessary.

Similar problems can exist in buildings used for some industrial processes, such as brewing and dyeing. Each case should be examined carefully and a design established in the light of a real understanding of the particular dangers involved.

THE HEATING SYSTEM

Sufficient has already been written about the need for maintaining general heating levels so as to give comfortable conditions and reduce the risk of condensation. In the average office, shop or school building condensation is not a problem, because these criteria are automatically observed and, volume for volume, high humidity areas in these types of building are insignificant. The building type most likely to suffer condensation, because of insufficient heating, is the domestic building, and in particular those parts of the domestic building which are only intermittently occupied – the bedrooms.

Heating is an expensive part of any family's budget – for old-age pensioners, or for persons on a fixed income, it can be crippling. This, therefore, is bound to be an area of compulsory family economy, often leading to the ineffective use of the heating system, or 'topping-up' by using cheap-to-run paraffin heaters, which aggravate the condensation risk. There is certainly a case to be made for old persons' dwellings being provided with landlord-controlled heating, even if it is only background heating. A few years ago this demand was being voiced for all local authority housing, but with the present financial climate this is unlikely to come about, except maybe in those places where a district heating scheme can be designed as an inexpensive, bonus by-product of an industrial process. There is no real alternative to proper heating if condensation is to be avoided. Mechanical dehumidification would work, but it is very expensive – and still the dwelling has to be heated for reasons of comfort.

The advice is self-evident. Minimum comfort temperature levels (Chapter 2) should always be maintained. These should be associated with reasonable ventilation and a structure that is thermally efficient. Then the risk of condensation is removed. The solution is in the hands of the designer. He has to place inside his thermally efficient shell a heating system that can produce these conditions and is inexpensive to run. The objective should be a general air temperature of about 15°C and the system should ideally run throughout the night – certainly it should never allow the surface temperatures of the structure to fall below 10°C. As stated before, a steady background temperature, even at a relatively low level, is preferable to short, sharp bursts of heat punctuated by long periods with no heat input.

There is a strong argument for those dwellings which will be occupied all day – old people's dwellings for instance – not being equipped with quick reaction, expensive-to-run systems, such as some warm air systems. These dwellings should be provided with continuous systems, complemented by a heavyweight building structure to take advantage of its thermal capacity. A hot water radiator system would seem to be the most effective in these circumstances. In the case of intermittently occupied dwellings, a quick reaction warm air system and a lightweight structure would seem to present the best solution, provided that a reasonable minimum temperature is always maintained and that, if the warm air unit is sited in the kitchen, it does not distribute the kitchen atmosphere around the whole house. Small source radiant heaters (as with warm air systems infrequently used) do little to heat up the structure and should be considered a less than adequate method of heating, unless used to 'top up' a generally good level of background heating. Both low temperature radiant floor and ceiling heating systems have advantages in controlling condensation. They warm up the structure, but they are only really effective when used in heavyweight structures.

One final point in the choice of a heating system: those systems which are flueless – electrical systems, or gas balanced flue and room sealed systems, or heating serviced by group or district heating arrangements – all lack the strong flue draughts that induce a certain amount of household ventilation. Maybe additional, permanent forms of ventilation should be provided in dwellings with flueless systems to compensate for this. The previous remarks about bedroom ventilation would also apply.

THE ENCLOSING ELEMENTS OF THE BUILDING SHELL

Surface condensation

In the previous chapter it was demonstrated how the risk of surface condensation can be calculated. What this showed, in effect, was that if the thermal resistance of the enclosing element was high enough, surface condensation would not occur, given a reasonable inside temperature. In fact if the thermal insulation levels required by the Building Regulations (see Appendix 1) were met, it would be unlikely that surface condensation would occur in most normal building types. This applies equally to walls or roofs, and even to ground floors, whether suspended or next to the ground. With reference to the latter, it is essential for perimeter insulation always to be included (see Chapter 5) in order to avoid surface condensation. A number of different enclosing elements and their thermal transmittancies are included in Table 5.10 and these will act as a guide to the type of structure that should be contemplated in order to provide varying levels of thermal performance. Also included in this table is the expected difference between surface temperature and air temperature, given stated internal and external climates; from this the danger of surface condensation can be judged against varying relative humidities (RH) using the graph in Figure 6.1.

Condensation on the glass in windows is a perfect example of surface condensation, and on single glazing, at any rate, is unavoidable. Much surface condensation is deposited on permeable surfaces, such as plaster, and consequently some of the moisture soaks into the surface material. In the case of glass (or metal window frames) the whole of the moisture deposited is unabsorbed and is therefore obvious. If the internal surface temperature of the glass is below the dew point of the internal air condensation will result, and glass, being a good conductor of heat, quickly assumes a temperature not far above that of the external air. Condensation on single glazing is, therefore, frequent.

Such condensation can be annoying, not only because it restricts visibility, but also because it can cause deterioration of the cill and surrounding surfaces. In addition, condensation can appear on the window frames, particularly if they are of metal, are uninsulated and without a thermal break (Figure 5.9). Generally, however, if there is a means of allowing this condensation to drain away, it is nothing more than a nuisance.

Even if double, or treble, glazing is used there is the possibility that at some times condensation will occur. If we assume an internal temperature of 12°C, an outside temperature of 0°C and an internal dew point temperature of 6°C – typical of normal bedroom conditions – condensation would take place on single glazing, but not on multiple glazing. Even if the temperature inside the building were raised to 20°C with the same humidity, there would still be condensation on single glazing. With the humidities that could be expected in the domestic kitchen, however, condensation would appear even on double glazing. If the RH, indoor and outdoor temperatures are known, it is possible to predict the incidence of condensation on various glazing systems from the graph in Figure 7.5. The thermal transmittances for the glazing systems should be compared with those given in Table 5.14.

As an example of the use of this graph, let us assume a glazing system with a U value of 5.6 W/m²°C (single glazing) and an internal RH of 60% (point A on the graph). If a line is drawn horizontally from this point to the inside temperature of 15°C (point B), it will be seen that condensation would form on the glass when the external temperature fell to about 3°C. If the external temperature fell to 0°C, a glazing system with a U value of about 4.3 W/m² would be necessary to avoid condensation (point C on the graph). This is the equivalent of a double glazed system with a 3 mm air space between the sheets of glass. If the inside temperature were raised to 20°C at the same RH and with the same outside temperature, the U value of the glazing system required would drop to about 3.5 W/m²°C (point D) or a double glazed system with a 12 mm air space between panes.

It has been suggested that in the case of the average domestic window-to-wall ratio, double glazing is not worth the capital cost. The saving in fuel bills is relatively small and the recouping of the initial cost prolonged. In addition, the BRE suggest[1] that in all-up fuel saving terms double glazing is hardly worth the expenditure. 'It is only in the most favourable case, that of a new dwelling heated to a high standard and with good controls, that the measure is just about cost-effective, and even then only at the severest cost profile.' The situation is obviously different when the window-to-wall ratio is considerably higher than that met with in the average dwelling.

The question of ultimate worth can equally well apply to that type of double glazing in which a secondary aluminium frame is fixed to the inside of the existing

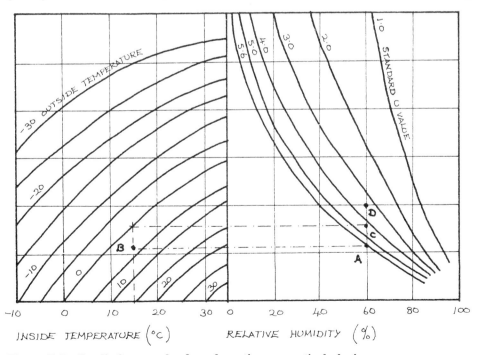

Figure 7.5 Prediction graph of condensation on vertical glazing

timber window. These are thermally effective only if the inside glazing is as air-tight as possible; if there is leakage, condensation can form between the panes of glass. There is also a substantial possibility that the existing timber window frames will rot before the capital cost of the aluminium system has been recovered.

The problem of condensation forming between panes of multi-glazed systems can be totally avoided only if sealed multi-glazed units are specified. If individual sheets of glass are used, ventilation needs to be provided to the cavity between the outside and the next layers of glass (Figure 7.6). This will slightly reduce the thermal resistance of the glazing system but, so long as the ventilation is minimal, the reduction will be hardly significant.

It can also be argued that the formation of condensation on a single glazed window is a form of insurance policy – the glass acting as a dehumidifier, removing from the atmosphere water that would otherwise condense elsewhere with more dire consequences. This point of view, however, is valid only if the single glazed windows are provided with condensation channels along their cills to collect the run-off from the glazing, which is then drained harmlessly to the outside through 6 mm or 9 mm diameter plastic or copper tubes built in through the cill at 300 mm centres (Figure 7.7). This is particularly important when the windows are large in area.

Interstitial condensation   It was explained in the previous chapter that the risk of interstitial condensation can be calculated by comparing the temperature gradient through an enclosing element with the dew point gradient, achieved by examining the effect of the vapour resistances of the various layers of the element and the actual vapour passing through to the various layers of the wall. When the temperature gradient falls below the dew point gradient, interstitial condensation will form. It was also pointed out that this can result in the danger of rot in vulnerable materials of construction, or reduced thermal efficiency of the insulants. From this the principle of the use of vapour checks or the ventilation of some internal cavities was established.

Vapour checks (sometimes referred to as vapour barriers) comprise a layer or membrane of high vapour resistance, such as a plastic sheet, bitumen felt, metal, metal foil, or even gloss paint. A few common examples are given in the table of vapour resistivities (Table 6.2).

The term 'vapour check' is used generally in this book, as opposed to 'vapour barrier', because almost all vapour barriers are difficult to install with assurance of

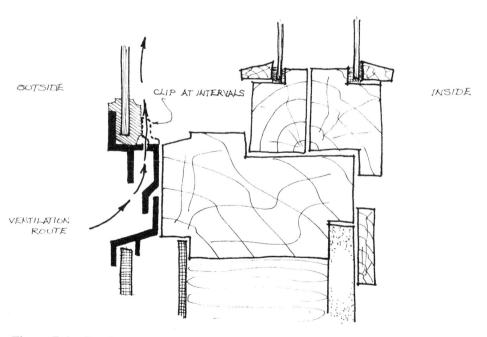

Figure 7.6 Continental multi-glazed system incorporating ventilation

their total effectiveness. Placed behind wall or ceiling linings they tend to be punctured by pipes and electrical points; also the sealing of barriers at wall-to-ceiling and wall-to-wall junctions is sometimes difficult to achieve. It is therefore prudent to consider that all barriers of this type will function at only a low rate of efficiency – maybe 50% of the declared resistance of the material. Hence the *barrier* becomes a *check*, a substantial deterrent to vapour passage but not an entire elimination of the risk. If, in calculation, it can be seen that condensation risk can be avoided with only 50% efficiency from the vapour check, there is reasonable hope that the situation in practice will be satisfactory. In questions of condensation it is always advisable to err on the safe side. The only type of vapour barrier that can be considered a fully effective barrier is one such as a bitumen felt barrier laid in bitumen on a flat roof deck. Here the membrane is rarely punctured and, if it is, sealing round a punctured membrane is simple with the bitumen used in its laying. The same comment applies to the sealing of the lapped joints between adjoining sheets of felt. But even in this almost foolproof example, especial care is needed to ensure complete sealing at junctions of the roof with up-stands or parapet walls.

The effectiveness of the jointing between adjoining sheets of vapour check material is obviously critical to the performance of the check. Figure 7.8 shows some of the joints which are to be preferred. Only the first example can be considered as a true vapour barrier – the sealed lapped joint in a bitumen felt membrane. The other examples refer to vapour checks.

Aluminium-foil-backed or polythene-backed plasterboard is often used to produce a vapour check. If successful this is clearly a cost-effective solution, achieving two functions with one fixing operation. But the sealing of the joints between plasterboard sheets can be critical, although recent research by the BRE has suggested that holes in the check (for instance where switch boxes occur) can be more harmful to the proper functioning of the vapour check than the plasterboard joints, assuming they are both well filled and backed by a timber framing member.

Recent experiments at the BRE have discovered some interesting facts about the effect of plasterboard, both plain and backed, on the vapour permeation into an adjacent cavity. For instance, it has been shown that if vapour pressure alone operates it takes several hours for the cavity to reach equilibrium with changed room

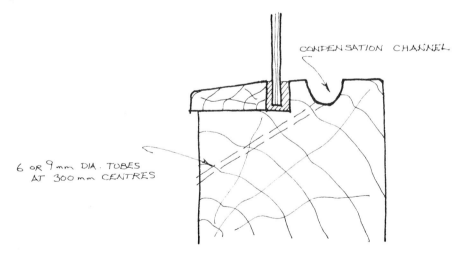

Figure 7.7 Condensation channel and drainage tubes

humidity either through holes in the plasterboard or through open joints. If, however, a positive air pressure of 20 N/m² is imposed across the wall lining (the standard test pressure applied in Denmark in interstitial condensation tests) equilibrium occurs in a few minutes. It is possible that this type of pressure differential is excessive for the average domestic building – 10 N/m² would seem to be more realistic. In fact, the BRE research has shown that, generally, plasterboard is more slow to react to atmospheric humidity changes than had been previously believed. Even plain plasterboard, without holes, will protect a wall cavity from changes in room conditions for several hours, because of its ability to absorb the water vapour. It has also been shown that this ability to absorb moisture can help to dry out a damp cavity which has suffered from short-term condensation, the lining re-absorbing the moisture through its unprotected back surface. The plasterboard is, in effect, acting as a balancing agent for the moisture in the cavity. Clearly this ability to re-absorb moisture from the cavity would be lost if the plasterboard were backed. This has led to some authorities suggesting that when periods of high humidity in buildings are likely to be of short duration, plain plasterboard linings are to be preferred, provided that the external linings of the cavity are permeable and allow the dissipation of the water vapour to the outside.

This is, however, a dangerous viewpoint and one not generally held, particularly in respect of low-income group housing. Even damaged plasterboard backings produce very much greater resistance to vapour passage than plain plasterboard and, particularly when timber framed construction is involved, the risks of condensation causing rot must never be under-estimated.

The conclusions, then, must be that backed plasterboard with sealed holes and joints is normally to be recommended when there is danger of structural or insulation damage; but, due to the likelihood of damage, only a percentage of the backing's vapour resistance should be assumed and the outer sheathing of the wall elements should be permeable (in this respect bitumen impregnated fibreboard would be preferable to exterior quality plywood), or the cavities should be ventilated.

Vapour checks should always be installed on the warm side of the the insulation so that there is no danger of condensation forming on the inside surface of the check. It is therefore normal to position the vapour check near the inside surface of the enclosing element, immediately backed by the insulation. This order would be reversed in the case of air-conditioned buildings in hot and humid climates. As indicated before, it is usually unnecessary to use vapour checks in heavyweight

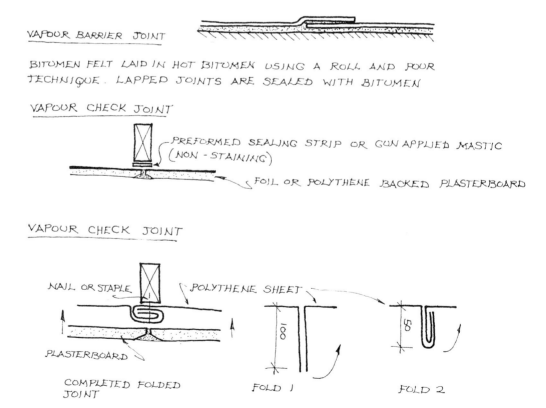

Figure 7.8 Vapour barrier/check jointing

structures where there is no danger of deterioration if interstitial condensation does occur; but in lightweight structures they are always essential. In addition, in structures with an unintentional vapour check towards the exterior of the enclosing element (metal cladding, or an asphalt or bitumen-felt roofing layer) the importance of an effective check cannot be over-emphasised, as trapped water vapour between checks can cause an accumulation of interstitial moisture. If it is possible to ventilate the space behind such an external layer (usually no problem in the case of wall cladding) any vapour build-up can be dissipated. This is not so easy in the case of a roof. When tile hanging or weatherboarding is the exterior cladding, any membrane behind the cladding should allow the passage of water vapour.

Vapour checks in walls

Table 5.10 shows a number of common wall constructions and gives advice on the necessity of vapour checks in each case. Normally in brick or block constructions a vapour check is not necessary, unless there are large quantities of vapour present. It is always advisable, for instance, to treat the surfaces of high-humidity domestic areas with a vapour resistant type of decoration – gloss paint, or vinyl paper – to discourage the vapour passing into the surrounding structure. If, however, it is necessary to increase the thermal performance of heavyweight walls by the addition of linings, a vapour check is always necessary.

In timber-framed walls a vapour check is essential, particularly if the wall is clad with asbestos-cement, metal sheeting, painted timber boarding, or plastic weatherboarding – all of which inhibit the passage of water vapour. Tile hanging and unpainted weatherboarding will allow the wall to breathe, but even in these constructions a vapour check is advisable. The same recommendation applies to timber-framed walls clad in brickwork.

A breather paper should be installed behind tile hanging. This serves as an additional protection against driven rain and snow but does not prohibit the outward passage of water vapour. The bitumen impregnated breather paper should be hung horizontally with 100 mm straight-lapped joints. The space between the breather paper and the tile hanging should be ventilated.

In steel-framed structures a vapour check should also be used. Cold bridging (see Chapter 5) is a common fault with any type of metal framing and particular care needs to be taken to include sufficient insulation at framing positions to avoid a great differential between the U values of different sections of the same wall. If a differential occurs it is extremely likely that differential staining (pattern staining) will occur, and if the effect of the cold bridge is to reduce the surface temperature of the wall at that point below air dew point, surface condensation will also occur.

Vapour checks in roofs

Roof cavities without a vapour check at ceiling level can be subject to intense condensation. Water vapour passes through the ceiling and condenses out on the first cold surface it meets, usually the underside of the sarking felt or boarding of the pitched roof, or the decking or roofing membrane of flat roofs. This usually means in a pitched roof that the vapour has passed right through the insulation, often located immediately above the ceiling lining. If there is insufficient air movement in the roof space to dry out the condensation, the water can drip back on to the insulation, thereby eventually degrading its performance. The amount of moisture involved can cause long-term high moisture contents in the roof timbers. If this is in excess of 25% it can mean, eventually, the risk of rot and even structural collapse. In flat roofs, depending on the construction and the position of the insulation, condensation can either degrade the insulant, endanger the roof timbers, cause the roof membrane to tear due to excessive moisture movement in a timber roof deck, or produce blisters of water vapour under the roof membrane which will eventually cause its premature breakdown. There is also the danger, either in flat or pitched roofs, of the dripping of water on to to the ceiling lining causing mould growth and staining on the underside of the ceiling.

In pitched roofs of whatever pitch a vapour check at ceiling level is strongly recommended. There are varying opinions on this point, some authorities maintaining that in the case of roofs over 20° pitch, a vapour check is unnecessary due to there being sufficient air movement in tall roof spaces to dry out any condensation that may occur. This view is not upheld by the evidence of many cases of extreme condensation in such roofs, particularly if the sealing of the roof space against vapour from the kitchen and bathroom is not adequate, and when impermeable sarking is used (plastic sheeting).

The vapour check should preferably take the form of polythene sheeting fixed independently to the ceiling joists before the ceiling is lined. This allows the jointing of the polythene sheets to be effectively carried out (Figure 7.8). Backed plasterboard is more usually installed, but the doubts about its effectiveness when used on walls are here accentuated. The habit on some low-cost buildings of not backing the end joints of the plasterboard with noggings makes the efficiency of the check even more doubtful. Insulation should be laid immediately above the vapour check.

Pitched roof spaces should be ventilated along the opposite eaves with the equivalent free area of a continuous gap of 10 mm in the soffits of roofs above 15° and 25 mm below 15°. It is sometimes argued that in particularly humid parts of the UK (Scotland is usually mentioned in this context) such ventilation causes moisture-laden air to be sucked into the roof space, thereby aggravating, rather than alleviating, the problem. If, however, it is decided to omit the ventilation, the vapour

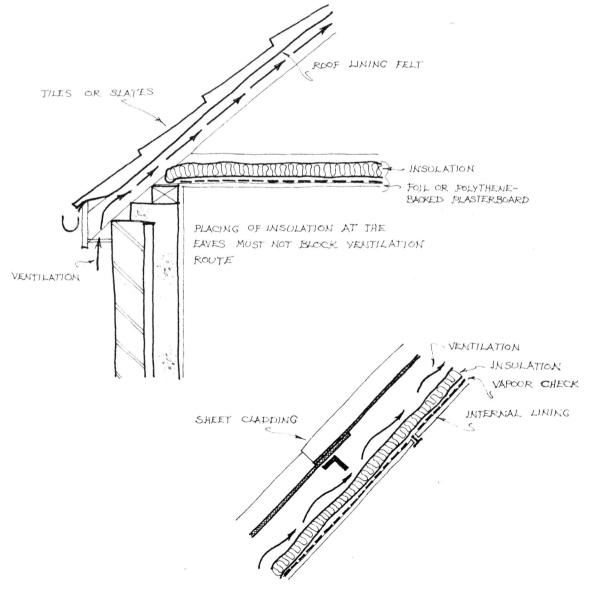

Figure 7.9 Typical pitched roof constructions showing ventilation routes and vapour checks

check must be installed with special care in an effort to make it as nearly perfect as possible.

Pitched roofs clad with asbestos-cement or metal sheeting present a particularly critical condensation risk, due to their dramatic temperature drop on clear, frosty nights (see Chapter 6). It is usual to line the inside face of these elements with insulation board or plasterboard, supporting a quilt of insulation. It is essential that a vapour check be provided, either by the use of a backed plasterboard (sometimes plastic-faced) or by finishing the interior surface with a sprayed plastics coating. None of these expedients is entirely effective and so it is advisable that the space behind the cladding should be ventilated. This can prove difficult at times due to the Building Regulation (Part E) requirement for cavity barriers.

ᵣ Figure 7.9 shows typical roof constructions and the recommended treatment of the vapour checks.

Timber flat roofs can be either of the 'warm' or 'cold' form of construction, depending on whether the insulation is placed above or below the roof cavity. No attempt should be made to combine these two types of construction.

The cold roof, similar to the pitched roof, has insulation at ceiling level and a vapour check below the insulation. The space above the insulation must always be ventilated with at least 1000 mm² free area per linear metre length of eaves on two opposite sides of the roof. This requirement for ventilation can often prove difficult to satisfy when flush eaves or parapets are used. The Scottish custom of solid strutting between joists is a further complication, as are the Building Regulation requirements for cavity barriers.

Cold roofs are cheaper to construct than warm roofs, but condensation in them tends to be somewhat unpredictable and therefore the warm roof is to be recommended. The insulation in this type of construction is placed above the roof decking and immediately below the roofing membrane with a vapour barrier between the decking and the insulation. The ceiling should have a low thermal insulation value to allow the roof cavity to be kept warm. It must also be permeable.

Concrete flat roofs should always be constructed on the warm roof principle. When wet insulating screeds are used, care must be taken not to trap construction water in the screed. A vapour check should never be installed under a wet screed, and ventilation paths should be supplied to allow drying out to take place (Figure 7.10). Wet screed over dry insulation is not recommended even if there is a vapour barrier between them. The screeds are likely to crack and curl.

Dry insulation is the better solution, placed above either the screed or the structural concrete. Rot-proof insulation (fibre glass or expanded plastics) should be specified. A vapour barrier laid in bitumen below the insulation is essential.

Vapour barriers of bitumen felt should be bonded to concrete surfaces with the 'pour and roll' technique. When treating decks of high moisture movement, a sealing strip of felt should be laid to the deck on the line of the felt lap joints prior to the main membrane being laid. This should avoid the possibility of the felt splitting at the later stage.

Figure 7.10 shows typical flat roof constructions, including the 'inverted' roof in which the insulation is placed above the roof membrane. In this construction the roof membrane performs a double duty, combining its weathering performance with that of a vapour barrier. This is a valid form of construction, provided that the insulation is of a type that will not degrade when exposed to moisture.

CONDENSATION ON PIPEWORK AND CISTERNS

One final area of condensation risk should be briefly mentioned. This is condensation on cold water pipes and cisterns. This can be merely a nuisance; it can also ruin decorations when the pipes are situated in a humid roof space and dripping from them causes staining on the ceilings. Cisterns placed in airing cupboards can cause similar dripping. If a drip tray cannot be installed under the cistern, the bottom of the tank should be covered with at least 12.5 mm of insulation. Pipework can either be treated with an application of anti-condensation paint, which might prove sufficient to stop the dripping, or preferably be covered with pre-formed insulation.

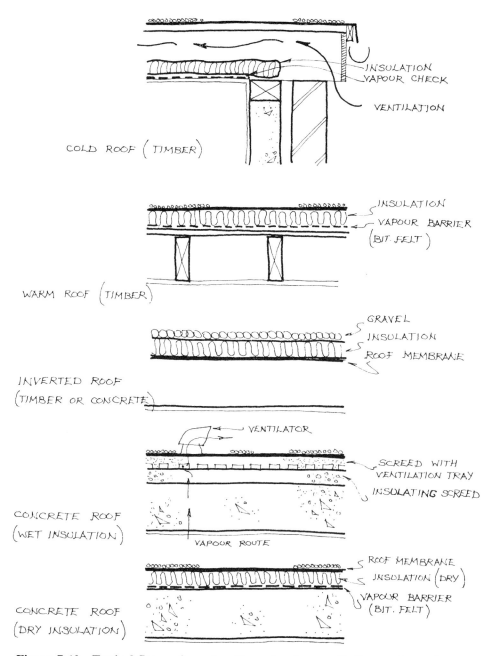

Figure 7.10 Typical flat roof constructions showing ventilation routes and vapour barriers and checks

BUILDING WATER

The average brick-constructed house contains about 1250 gallons of construction water. This moisture has to dry out, but may well take eighteen months to do so. Particularly this applies to next-to-earth ground floor slabs. Humidity levels will therefore be inflated during the first few months of occupancy of a new property. During this time the process of drying out could add about 2 kg of moisture to the internal atmosphere per day – roughly this is equivalent to two more occupants of the house breathing and indulging in sedentary occupations. Contractors can speed this initial drying out period by not allowing materials to become soaked by rain when stored on the site, enclosing the dwelling as soon as possible and completing the wet trades early in the construction programme.

REFERENCES

1. BRE, *CP56/75 A BRE Working Party Report: Energy conservation: a study of energy consumption in buildings and possible means of saving energy in housing*; BRE, 1975.
2. Rodwell, DFG, BRE Note *Water vapour diffusion through plasterboard linings on timber-framed walls*, BRE, 1976.

8 Remedial measures in existing premises

Before remedial measures can be taken, it is essential to diagnose accurately the true cause of the defect. Often the appearance of dampness on the inside surface of the building shell can be the result of rain penetration. (Conversely, staining caused by condensation is often initially interpreted as rainwater ingress). It is, therefore, always wise to endeavour to eliminate the possibility of dampness being the result of dirty wall ties, or blocked cavities, fractured or non-existent flashings, ineffective damp-proof courses or membranes, fractured roof membranes or coverings, etc., before jumping to the conclusion that the source of the trouble is condensation.

There will be occasions when it is perfectly clear that the problem is one of condensation – in buildings of high internal humidity, poor insulation, or poor heating. There will also be the evidence which particularly relates to condensation – mould growth in cupboards on external walls, or behind furniture, in unheated, unoccupied bedrooms etc. There will be a number of instances such as those concerning staining on ceilings below roof voids, or high moisture contents in sole plates in timber framed walls, where the cause is by no means so obvious. A completely objective approach to the investigation is essential.

Whatever remedial treatment is suggested, it will always be based on three principles:

1. Reduction of the internal humidity.
2. Improvement of the thermal performance of the element experiencing the condensation.
3. Raising the heating levels.

In the list of defects and remedial measures that follow it is assumed that rain penetration has been eliminated as the cause of the defect and that, together with the treatment suggested, an effort is made to reduce, or better control, the humidity of the building in which the defect occurs.

DAMPNESS ON SOLID FLOORS NEXT TO EARTH

Cause

Lack of adequate perimeter insulation to the floor can lead to dampness. This defect is often mistaken for rising damp caused by a defective damp-proof membrane. If moisture is observed to form more often around the perimeter of the building, rather than generally, condensation is likely to be the cause. The defect can remain unnoticed under carpets for a considerable time, resulting eventually in the rotting of the carpet and underfelt.

Remedy

1. Insulating existing solid floors is expensive and rarely justified. A period of about two years should be allowed to elapse in new buildings before treatment is

seriously contemplated. Solid ground floors next to earth take a long time to gain the expected surface temperature consistent with the heat input of the building. Also the drying out of construction moisture may be the cause of the dampness and this will need time to dissipate.

2. If the defect persists, the floor can be covered with cork tiling, hardwood blocks or softwood flooring on treated battens.
3. This defect is most likely in single storey dwellings in unoccupied bedrooms. Better heating and ventilation will help to alleviate the problem.

DAMPNESS ON SUSPENDED GROUND FLOORS

Cause

This defect usually occurs only when an impervious floor finish (eg PVC sheet) has been laid over an uninsulated floor.

Remedy

1. If the floor is timber, it is better if an impervious finish is not applied, due to the danger of water build-up within the floor caused by spillage and washing down. In time this build-up will cause rot. If an impervious finish is unavoidable, additional insulation between the joists in a timber floor, or under a concrete floor would be an asset, if it is practicable for it to be installed.

DAMPNESS ON WALLS

Causes

If dampness is localised, it could be due to cold bridging within the structure, reducing the U value of that part of the wall. This can occur at lintels, ring beams, concrete floors and roofs penetrating the wall element, balconies and canopies, external corners and on the line of panel joints.

If dampness is general, it could be caused by inadequate thermal insulation of the wall element.

Note: Dampness on the internal surface of framed walls could be due to a dangerous build-up of interstitial condensation. Its appearance on the internal surface may well be localised. Any surface dampness on framed wall interior surfaces should be fully investigated to ensure that it is not the visible result of interstitial condensation.

Remedy

1. In the case of general condensation the high humidity of the building should be decreased by improved ventilation or the isolation of the humid areas. Heating levels in the building generally should be raised. Additional insulation to the wall elements should be contemplated. This could either take the form of cavity insulation (see Appendix 2) or be achieved by lining the walls internally with plasterboard backed by a vapour check and insulation.
2. Localised condensation can be avoided by the addition of extra insulation in the areas of the cold bridge.
3. In either local or general condensation, the wall can be treated with a wallpaper on a 3 mm polystyrene sheet, or polystyrene tiles. The polystyrene should be of the type that discourages the passage of water vapour and the adhesive used should not be sensitive to water.

   The fire implications of any insulating material adhered to the wall face should be considered. Many plastic materials subjected to fire emit choking fumes even if they do not, in fact, burn.

   Note: The use of anti-condensation paint is unlikely to prove satisfactory. Anti-condensation paint merely discourages the condensation running on the surface by providing a porous outer layer to the wall. It does not stop moisture forming on the surface.

MOULD GROWTH ON WALLS AND CEILINGS

Cause
A high moisture level on the affected surface is the cause of mould growth.

Remedy

1. First the spores of the mould have to be destroyed. On undecorated surfaces apply a fungicide (common household bleach is often adequate – sodium hydrochlorite – but other proprietary fungicides are available). After a period brush down the surface, and wash with a small quantity of water if this is necessary to restore the appearance of the surface. The treatment should be repeated if there are any signs of new growth. On decorated surfaces the treatment will depend on the type of decoration. When the mould growth is slight and the surface is robust but sufficiently absorptive to be impregnated by the treatment, the method should be as described for undecorated walls. In other cases the decoration will have to be stripped before sterilising the area with a fungicide. Vinyl wall-coverings are relatively impermeable and prone to encourage mould growth beneath them, producing pink or black stains. Fungicidal adhesives should always be used.

   A list of available fungicides is given in BRE Digest 139.

2. Remedies 1 and 3 quoted for 'Dampness on walls' apply equally in this case.

CONDENSATION ON THE INSIDE SURFACE OF WINDOWS

Cause
On single glazing condensation is inevitable during periods of external low temperatures. Condensation can also occur on double glazing when the humidities inside the building are high and the external temperature is very low.

Remedy

1. Double glazing can be substituted for single glazing. This will dramatically decrease the condensation, but is unlikely to have an appreciable effect on heating costs. Methods include:
   (a) reglazing the window with sealed units. In this case care must be taken that the glazing rebates are adequate to receive sealed units.
   (b) fixing secondary glazing to the inside of the existing windows. This method tends to produce better sound insulation than reglazing with sealed units, but, unless the secondary glazing system is well sealed against the passage of humid air into the cavity between panes, condensation can form on the outside layer of glass. Between glazing condensation can encourage rot in timber window frames. Secondary glazing systems when attached to timber windows are likely to out-live the windows.
   (c) replacing the windows with new windows with double glazing.
   (d) the installation of the heat source at the base of the glass in large glazed areas can help to reduce condensation – and, incidentally, the cold down-draught.

2. The cill of timber single glazed windows can be grooved to form a condensation collection channel. Holes drilled from this through the cill and lined with 6 mm or 9 mm plastic or brass tubes will discharge the condensation outside.

3. There are certain proprietary films and fluids which, when spread on the inside of glazing, discourage the formation of discrete droplets of water and thereby aid clarity of vision. This method, however, will not cure condensation, it merely changes the misting over of the glass. Repeated window cleaning will result in the need for frequent renewal of the treatment. In some large scale applications in florists' shops the inside surface of the glass is covered with a smooth sheet of water flowing from a feed pipe at the top and collected in a trough at the bottom.

Note: A case can be made for accepting condensation on windows due to its dehumidifying effect on the internal atmosphere.

Between-glazing condensation, however, is a nuisance and can be eliminated by ventilating the air space between the glazing systems by drilling holes from the outside into the air space. This will marginally reduce the U value of the glazing system.

STAINING ON CEILINGS BELOW ROOF SPACES

Cause

Interstitial condensation in the roof space, dripping from either the sarking felt or from cold water storage cisterns or pipe work, can result in staining of the ceilings underneath.

Remedy

1. Improve the vapour resistivity of the ceiling by the addition of a plastic sheet type of vapour check immediately above the ceiling lining. First remove any insulation from the top of the ceiling lining. Lay plastic sheets, dressed down between the ceiling joists or the ties of the trusses. Adjoining sheets should have welted joints. In practice this can prove very difficult (in some forms of roof construction, impossible). For the remedial treatment to be effective the vapour check needs to be held as close as possible to the ceiling lining between the roof timbers, in order to avoid an air space. The insulation is next replaced above the vapour check. If no insulation existed in the roof previously, it should now be installed.

   Note: This defect is often the direct result of placing insulation in the roof space, thereby lowering its temperature but not its moisture content.

2. If (1) is not possible (and perhaps even if it is) increase the ventilation of the roof space. Recommendations are given in Chapter 7 for eaves ventilation. Make sure that the installation of insulation has not blocked up the air routes. Air bricks can also be placed in the gable walls, but 5000 mm² free area will need to be provided in each gable for every 1 m length of eaves – usually one 225 mm square air brick to every 1 m of gable length.

3. In addition to the treatment of cold water storage cisterns and pipe work as described later, cold water storage cisterns should be properly covered. It is often the practice not to place insulation under the cistern when it is located in a roof space. In this way freezing up is discouraged. It will, however, result in the water being kept warm enough to evaporate into the roof space atmosphere readily. The tendency of some central heating systems to 'pump over' steam or hot water into the expansion tank can also cause evaporation. Reducing the pump rate should stop this.

CONDENSATION DRIPPING FROM COLD WATER PIPEWORK

Cause

When warm, humid air coming into contact with cold pipes is so depressed in temperature that it can no longer contain all the water vapour it was holding, the excess will be deposited as condensation on the pipes.

Remedy

1. Painting the pipes with anti-condensation paint may be sufficient to avoid dripping if the condensation is not excessive.
2. Surround the pipes with preformed insulation.

CONDENSATION DRIPPING FROM COLD WATER STORAGE CISTERNS

Cause

Condensation forms on the cold surface of the cistern in the same way as on cold pipes.

Remedy

1. Install a drip tray below the cistern.
2. Place insulation at least 12.5 mm thick under the cistern. This may be unwise in excessively cold roof spaces. Complete insulation of the cistern may be advisable.

In choosing an appropriate type of insulation material, for whatever location, consideration should be given to any fire hazard that may attach to the particular material – not merely to its own flammability but also to whether it gives off dangerous fumes during a fire. Possible corrosion that may take place between the insulation and the nail plates of the roof trusses is another example of the side effects that have to be considered when choosing an insulant. The current edition of Insulation Handbook[1] is a useful reference work.

REFERENCE

1. *Insulation Handbook, 1978/79*, Lomax Wilmoth and Co Ltd.

Appendix 1 – Mandatory requirements

ENGLAND AND WALES

Regulating instruments

The Building Regulations 1976, Part F, Thermal Insulation; and Building (First Amendment) Regulation 1978 (operational from 1st June 1979) are the regulating instruments.

Requirements

The Building Regulations, Part F concerns dwellings (houses, flats or maisonettes); the Amendment Regulation adds Part FF to the regulations, entitled 'Conservation of fuel and power in buildings other than dwellings'. This amendment makes obsolete the provision of the Thermal Insulation (Industrial Buildings) Act 1957, which will be repealed for England and Wales (except Inner London).

Part F sets down minimum standards of insulation to main enclosing elements, *excluding* openings in the element.

Maximum U value (W/m² °C)	Elements
1.0	External wall, wall between dwelling and a ventilated space, wall to rooms in roof (including insulation of roof space and structure), wall next to roof space over a dwelling, external floor, floor next to ventilated space.
1.7	Wall between dwelling and partially ventilated space, wall next to a building not covered by Part F.
0.6	Roof

Table A1.1   Minimum standards of insulation

Part F further stipulates a maximum *average* U value for perimeter walling (1.8 W/m² °C) *including* openings and gives U values to be used in this average calculation for single and double glazing (5.7 and 2.8 W/m² °C).

This regulation gives lists of deemed-to-satisfy constructions.

Part FF refers to all other building purpose groups, except buildings not exceeding 30 m² and those whose heating is designed not to exceed 25 W per square metre of floor area (or, in the case of purpose groups VI (factories) and VIII (warehouses) when used solely for storage, 50 W per square metre).

It restricts glazed areas to percentages of wall and roof areas (Table A1.2).

	Purpose group		
	II or III	IV, V, VII	VI, VIII
Windows %	25	35	15
Rooflights %	20	20	20

Table A1.2   Restriction of glazed areas

Maximum U value (W/m² °C)		Elements
Purpose groups II, III, IV, V, VII, VIII (not wholly storage)	VI and VIII (when used for storage)	
0.6	0.7	External wall, internal wall next to ventilated space, floor exposed or next to ventilated space, roof (other than over ventilated space or partially heated space).

Table A1.3   Maximum U values

or the heat loss equivalent to that through single glazing of the specified areas, and also sets down maximum U values for elements of enclosure (Table A1.3).
Part FF also includes a list of deemed-to-satisfy constructions.

SCOTLAND

Regulating instrument

The Building Standards (Scotland) Amendment Regulations 1975 is the regulating instrument.

Requirements

These regulations are broadly in line with Part F of the English Regulations, and therefore deal only with dwellings. At the time of writing there is no equivalent to Part FF, although it seems likely that this will eventually be devised.

An additional clause (J6) is included in the 1975 Amendment and deals with interstitial condensation, requiring that the designer should 'design the structure so as to prevent, so far as is reasonably practicable, damage to any part of the building as a result of the passage of moisture in the form of vapour from the interior of the building into its structure.'

SCANDINAVIA

Regulating instrument

Riktlinger för Värmeisoleringsbestämmelser (Guide lines for thermal insulation regulations) NK Skrift Nr 8 (1967) is the regulating instrument. This is a model regulation that has been adopted in the Swedish Regulations.

Requirements

The regulation deals with buildings for 'constant occupation' – habitable rooms, with minor modifications for work spaces – and gives maximum thermal transmittances in kcal/m²h°C (1 kcal/h = 1.163 W) in four climatic zones for external walls, floors, roofs and windows (related to three ratios of fenestration to wall surface).
Reference is made to the avoidance of condensation and draughts.

WESTERN GERMANY

Regulating instruments

The regulating instruments are: German standard DIN 4108 Wärmeschutz Hochbau (Thermal insulation in buildings) 1969; DIN 4701 Regeln fur die Berechnung des Wärmsbedarfs von Gebaüden (Rules for the calculation of heating requirements for buildings); DIN 52612 giving methods of determining thermal conductivity.

Requirements

DIN 4108 applies to all buildings 'for constant occupation' and requires three grades of insulation to enclosing elements relating to three geographical zones. Lightweight structures (less than 200 kg/m² are to have storage heating or constant central heating. DIN 4701 gives design temperatures in residential, commercial and school buildings (20°C is the normal room temperature required for sedentary areas). Double glazed windows are required in lightweight structures. Cold bridges are inadmissible. Methods are given for calculating thermal conductivity (Λ) in kcal/m²h°C and deemed-to-satisfy constructions are listed.

The need for vapour barriers and draughtproofing under certain circumstances is noted.

FRANCE

Regulating instruments

The regulating instruments are as follows: (1) CSTB (Centre Scientifique et Technique du Bâtiment) Notice technique NT 1 – 12 – 58, section V – April 1970, (2) Décret No 69–596 of 14th June 1969. Règles générales de construction des bâtiments d'habitation (General building requirements for dwellings) articles 6 and 8, (3) Clauses techniques générales (CTG) des organismes HLM (General technical rules covering building for members of the HLM – Habitations à loyer modéré – organisation, which builds dwellings for letting at subsidised rates).

Requirements

These instruments refer to housing. (1) is a guide to the implementation of (2) with deemed-to-satify insulation values. (2) requires structure and insulation to be such that an internal temperature of 18°C can be maintained and for ventilation to be sufficient to avoid condensation. (3) gives maximum values for the volumetric heat loss (linked to three climatic zones) of the enclosing structure measured in W/°C (also in kcal/h°C) per m³. Double glazing is recommended when glazing exceeds a stated proportion of the floor area. Recommendations are also given for the avoidance of condensation.

CANADA

Regulating instrument

The regulating instrument is the National Building Code of Canada 1965
Subsection 3.65 Minimum indoor temperatures
Section 4.7 Cladding
Supplement 5 Residential standards
Supplement 1 Climatic information

Various Canadian standards and government specifications for insulating materials are quoted, together with the ASHRAE (American Society of Heating Engineers) Guide for methods of calculation.

Requirements

23 occupancy types are defined to establish minimum design temperatures varying from 72°F for dwellings to 65°F for bowling alleys. Supplement 5 deals with dwellings. Three climatic zones are defined, and minimum overall thermal resistances are given for the enclosing elements, excluding doors and windows. Detailed requirements are also given for the installation of insulation and vapour barriers.

Appendix 2 – Cavity insulation: advantages and hazards

While the installation of insulation in the cavities of brick or block cavity walls is clearly an effective way of achieving high insulation values, particularly in existing buildings where the only alternative is dry-lining (with all the consequent cost and disruption of use), the primary purpose of the cavity must not be ignored. A solid 225 mm brick wall is likely to be subject to rain penetration. The same thickness of brick in two leaves, separated by a cavity, will be in no danger of penetration, assuming the cavities are clean and the wall is well constructed. By filling the cavity with any material, however resistant to moisture penetration, the risks of dampness will be substantially increased – particularly if accidental cavities in the filling occur, thereby producing easy bridging routes by which water, flowing down the back of the outer leaf or brickwork, can cross the cavity. Additionally, urea formaldehyde foam (one of the materials used specially for the treatment of existing buildings) is subject to shrinkage. This can result in the formation of accidental gaps.

The Agrément Board has performed much research into this problem and has become the chief authority on the subject in the UK. Only materials covered by an Agrément Certificate should be specified, and the Board's recommendations for its use should be followed. The Board's comments on the use of various types of cavity fill are dependent on the exposure of the site and the establishment of this exposure is based on the driving-rain index (BRE Digest 127). The rather crude area designations of the Digest have now been modified to take account of local geographical and topographical characteristics (Lacey, RE: Driving Rain Index, HMSO 1976 and the Agrément Board Information Sheet No 10: Method of assessing the exposure of buildings for urea formaldehyde cavity wall insulation, 1977).

At present the types of cavity insulation which have received Agrément Board approval fall into two categories: treatments that can be built into the wall at the time of construction and those that can be installed by injection at a later date.

IN-BUILT TREATMENTS

These treatments comprise rock wool fibre or expanded polystyrene slabs. The Agrément Board does not propose any exposure limitations on the use of the rock wool fibre slabs.

Current Agrément Certificates for rock wool fibre slabs consist of:
Certificate 75/292 Rockwool slab
Certificate 75/293 Rocksil slab

The Agrément Board proposes exposure limitations on the polystyrene slabs unless the cavity can be maintained at 50 mm or more. The cavity should never be less than 25 mm after the installation of the slabs (not a requirement in the case of rock wool fibre slabs). The following exposure limitations should be observed:

Sheltered areas; satisfactory for use in facing brick walls or rendered block walls.

Moderate and severe areas; satisfactory for use in *rendered* brick or block walls.

The appropriate certificate is:

Certificate 75/321 Jablite expanded polystyrene slabs (Amended 77/AM 25)

INJECTED TREATMENTS

With the exception of two Agrément Certificates, this type of treatment is restricted to urea formaldehyde foam. The exceptions are:

Certificate 75/307 Rentokil Rockwool
      – no exposure limitations.

Certificate 77/472 Thermobead (expanded polystyrene pellets)
      – no specific exposure limitations, but there is in the Certificate a warning of danger if the brickwork is not good.

Of the remaining urea formaldehyde certified treatments, all have the following exposure limitations:

Sheltered areas; satisfactory for use in facing brick walls, or rendered block walls of 1, 2, or 3 storeys.

Moderate areas with built-up low obstructions; as recommended for sheltered areas but only up to 2 storeys in height.

Moderate areas with greater exposure; satisfactory for use in rendered brick or block walls of 1 or 2 storeys.

Severe areas; satisfactory for use in rendered or otherwise protected brick or block walls of 1 or 2 storeys.

Certified urea formaldehyde treatments include:

Certificate No.	Product	Certificate No.	Product
75/268	Cosywall	75/322	Airfoam
75/272	Foamair	75/324	Allwarm
75/274	Ciba-Geigy	75/325	Heat Save
75/278	Cosyfoam		(Amended
75/282	Maxi Heat		76/AM 23)
75/284	Warmlife	75/326	Thermawal
75/302	Saxonfoam	75/327	Eurofoam
75/303	Foamalon	75/328	NIS
75/304	Interfoam	75/329	Hawkins
75/310	Phoenix	75/331	Polymer
75/311	Thermafoam	75/333	Injectawarm
75/312	Insultreat	75/336	Kavitex
75/313	K-Nine	75/339	Wall Lag
75/314	Foamatherm	76/347	Fulfoam
75/315	Cosyfoam	76/373	MPI foam
75/316	Econoseal	76/394	Insuwall
75/317	Thermoseal	76/395	Isofill

Appendix 3 –
List of advisory bodies

Agrément Board,
Lord Alexander House,
Hemel Hempstead, Herts. HP1 1DH

British Standards Institution,
British Standards House,
2 Park Street,
London W1A 2BS

Building Research Establishment,
  Building Research Station,
  Garston, Watford WD2 7JR

  Princes Risborough Laboratory,
  Princes Risborough,
  Aylesbury, Buckinghamshire HP17 9PX.

  Building Research Establishment – Scottish Laboratory,
  Kelvin Road,
  East Kilbride, Glasgow G75 0RZ.

Chartered Institute of Building Services,
49 Cadogan Square,
London SW1X 0JB.

Insulation Glazing Association,
6 Mount Row,
London W1Y 6DY.

Meteorological Office, advisory offices:

  Meteorological Office,
  Met 0.3 (b),
  London Road,
  Bracknell RG12 2SZ.

  Meteorological Office,
  26 Palmerston Place,
  Edinburgh EH12 5AN.

  Meteorological Office,
  Tyrone House,
  Ormeau Avenue,
  Belfast BT2 8HH.

USA

American Society for Testing and Materials,
1916 Race Street,
Philadelphia PA 19103

International Conference of Building Officials,
Whittier,
California.

Appendix 4 – Schedule of relevant codes of practice and standards

BS CODES OF PRACTICE:

CP 3

Chapter 11: 1970 Code of basic data for the design of buildings: Thermal insulation in relation to the control of environment.

Chapter V: Part 2: 1972 Code of basic data for the design of buildings: Loadings.

Chapter VIII: 1949 Code of functional requirements of buildings: Heating and thermal insulation.

BRITISH STANDARDS:

BS 3837: 1977	Expanded polystyrene boards
BS 3869: 1965	Rigid expanded polyvinyl chloride for thermal insulation purposes and building applications.
BS 3958:	
Part 1: 1970	85% magnesia preformed insulation
Part 2: 1970	Calcium silicate preformed insulation
Part 3: 1967	Metal mesh faced mineral wool mats and mattresses
Part 4: 1968	Bonded preformed mineral wool pipe sections
Part 5: 1969	Bonded mineral wool slabs (for use at temperatures above 5°C)
Part 6: 1972	Finishing materials; hard setting compositions, self-setting cement and gypsum plaster.
BS 4046:	
Part 2: 1971	Compressed straw building slabs
BS 4841:	
Part 1: 1975	Rigid urethane foam for building applications
Part 2: 1975	Laminated board for use as a wall or ceiling insulation.
BS 5250: 1975	Code of basic data for the design of buildings: The control of condensation in dwellings.
BS 5368:	
Part 1: 1976	Air permeability tests

Appendix 5 – Further reading

ASTM Special Publication 552, *Window and wall testing*, ASTM, 1972
Building Research Establishment current papers:

CP79/68 Loudon, AG, *U Values in the 1970 guide*, 1970

CP2/74   Milbank, NO, *A new approach to predicting the thermal environment in buildings at the early design stage*, 1974

CP7/74   Petherbridge, P, *Limiting the temperature in naturally ventilated buildings in warm climates*, 1974

CP8/74   Petherbridge, P, *Data for the design of the thermal and visual environments in buildings in warm climates*, 1974

CP20/74 Whiteside, D, *Cavity insulation of walls: a case study*, 1974

CP61/74 Milbank, NO, and Harrington-Lynn, J, *Thermal response and the admittance procedure*, 1974

CP80/74 Humphreys, MA, *Environmental temperature and thermal comfort*, 1974

CP56/75 *BRE Working Party: Energy conservation: a study of energy consumption in buildings and possible means of saving energy in housing*, 1975

CP8/76   Bloomfield, DP, *The effect of intermittent heating on surface temperatures in rooms*, 1976

CP64/76 Courtney, RG, *Solar energy utilisation in the UK: Current research and future prospects*, 1976

CP75/76 Humphreys, MA, *Desirable temperatures in dwellings*, 1976

CP14/77 Bloomfield, DP, Fisk, DJ, *The optimisation of intermittent heat*, 1977

CP9/78   Humphreys, MA, *The optimum diameter of a globe thermometer for use indoors*, 1977

CP17/78 Humphreys, MA, *A study of the thermal comfort of primary schoolchildren in summer*, 1978

Building Research Establishment digests:

108 *Standard U values*
110 *Condensation*
119 *The assessment of wind loads*
139 *Control of lichens, mould and similar growths*
140 *Double glazing and double windows*
145 *Heat losses through ground floors*
190 *Heat losses from dwellings*
191 *Energy consumption and conservation in buildings*
210 *Principles of natural ventilation*

Building Research Establishment Notes:

Milbank, NO, *Energy consumption in tall office buildings,* 1974
Milbank, NO, *Energy consumption in 'other' buildings,* 1975
N101/76   Rodwell, DFG, *Water vapour diffusion through plasterboard linings on timber-framed walls,* 1976

Croome, DJ, and Sherratt, AFC, *Condensation in buildings.* The proceedings of the Conference held at York University, Jan. 1972, Applied Science Publishers Ltd 1972.

DOE, *Homes for today and tomorrow,* HMSO 1961

CIBS Guide, Book A, CIBS (until recently referred to as the *IHVE Guide*)

Insulation Handbook 1978–79, Lomax Wilmoth and Co. Ltd, 1978

Marsh, Paul, *Air and rain penetration of buildings,* Construction Press, 1977

MOPBW, *Condensation in dwellings, Part 1; a design guide,* HMSO, 1970

Pilkington Brothers Ltd, *Glass and windows bulletin No 1,* 1967
Solar control performance blinds, 1973
Solar heat gain through windows, 1974
Thermal transmission of windows, 1973

PSA Advisory Leaflets:
 34 *Thermal insulation,* HMSO, 1975
 61 *Condensation,* HMSO, 1967
 79 *Vapour barriers and vapour checks,* HMSO, 1976

Szokolay, SV, *Solar energy and building,* Architectural Press, 1975.

Appendix 6 – Glossary

admittance	A measure of the ability of a construction to admit heat, expressed in the same units as thermal transmittance (W/m² °C)
air temperature	The average temperature measured with an alcohol-in-glass or a mercury-in-glass thermometer at 1.5 m from the floor.
angle of altitude	The angle between the sun's rays falling on an object and the line between that object and the horizon – the angle of the sun above the horizon.
angle of azimuth	The angle of a vertical plane through the sun measured clockwise from the North.
angle of incidence	The angle at which the sun's rays strike an object – the angle measured between a line normal to a wall's surface and the sun's rays.
angle of wall- solar azimuth	The angle between a line perpendicular to a wall and the sun's azimuth.
ASTM	American Society for the Testing of Materials.
BRE	Building Research Establishment of the DOE, based at Garston, Watford, England.
BSI	British Standards Institution.
buoyancy	The natural tendency of air to rise when there is a temperature differential. The force that causes the *stack effect*.
cavity barrier	A barrier to the passage of smoke or flames in a constructional cavity. A requirement of Part E of the Building Regulations 1976.
CIBS	Chartered Institution of Building Services.
Clo	A unit of measurement of the insulation value of clothing. 1 Clo is roughly equivalent to the insulation of a normal suit (approximately 0.155 m²°C/W).
cold bridge	A path of more ready heat transmittance through a structure. A point of relatively high U value compared with the rest of the structure. Often this is caused by the bridging of a cavity by a structural component (synonymous with *warm bridge*).
'cold' roof	A roof structure with the insulating layer below the roof void.
comfort temperature	A measure of the effect of air and surface temperature of a room on the thermal comfort of a body occupying that room. In this book this is taken as the simplified *dry resultant temperature*.
condensation	The formation of water on a relatively cold surface which causes the adjacent air to be cooled below its dew point (see *interstitial* and *surface condensation*).
conduction	The transfer of heat through a solid material from a region of higher temperature to a region of lower temperature.
convection	The transfer of heat in or by a liquid or a gas by the movement of the medium.
cyclic conditions	The conditions present when heat in-put is fluctuating or intermittent.
decrement factor (f)	The ratio of the cyclic thermal transmittance to the steady state U value.
density	The mass per unit volume of material at a specific temperature measured in kg/m³.
dew point	The temperature at which a sample of moist air becomes saturated and at which condensation begins. It depends on the mass of moisture in the air and the temperature of the air.
differential pressure	The air pressure variation that can be experienced between different faces of a building or element of construction.

diffusance	The property of a specific thickness of a material which determines the rate of the passage of water vapour through unit area of the material at unit difference of water vapour pressure between surfaces (see *vapour diffusivity*).
diffuse radiation	In terms of solar radiation it is that part of the sun's radiation which does not depend on the direct incidence of solar rays on an object – the type of radiation experienced on overcast days.
diffusivity	The property of a material which is independent of thickness and is a measure of the rate at which vapour will pass through the material when a difference of pressure exists between the air on opposite sides. It is the reciprocal of *vapour resistivity* and is measured in gm/MN.
direct radiation	Radiation that passes in a straight line from the radiating object to the receiving object.
DOE	Department of the Environment.
dry resultant temperature	A method of measuring the combined effect of air temperature and mean radiant surface temperature of a room to arrive at a comfort temperature. Measurement is made by a thermometer enclosed in a blackened globe.
effective temperature	An arbitrary index of the degree of warmth or coldness felt by a human body in response to temperature, humidity and air movement.
emissivity	The ratio of the thermal radiation from unit area of a surface to the radiation from unit area of a full emitter (black body) at the same temperature.
environmental temperature	A simplified equivalent to the comfort temperature, combining air temperature with mean radiant surface temperature.
equivalent temperature	A method of measuring the combined effect of air temperature and mean radiant surface temperature in a room. It is defined in CP3: Chapter VIII: 1949 as 'the temperature of a uniform enclosure in which in still air a black cylinder of height 22″ and diameter 7.5″ would lose heat at the same rate as the environment under consideration, the surface of the cylinder being maintained at a temperature which is a precise function of the heat loss from the cylinder, and which in any uniform enclosure is lower than 100°F by ²/₃rd of the difference between 100°F and the temperature of that enclosure.'
global temperature	A method of measuring the combined effect of air temperature and mean radiant surface temperature of a room to arrive at a comfort temperature. It is measured using a thermometer inside a blackened globe.
insulation	See *thermal insulation*.
interstitial condensation	Condensation occurring within the thickness of a building element or within its component materials.
mean radiant temperature	The temperature of a black cylinder about 22″ in height and about 7.5″ in diameter in radiation equilibrium with the environment. The sum of the products of surface temperatures and surface areas taken over the whole room, divided by the surface area of the room.
orientation	The direction in which a building elevation faces in relation to the points of the compass.
partition factor	That proportion of the radiant heat absorbed by the glass of a window which is later released to the internal environment.
permeability	A property of a material which determines the rate at which water vapour passes through it under the influence of unit pressure. The *vapour diffusance* of a material divided by its thickness.
primary energy	The total energy before losses brought about by conversion into usable energy of various forms.
radiation	The process by which heat is emitted in rays from a hot source. Its operation is independent of air.
reflectivity	The ability of a surface to deflect and retransmit rays of light or heat incident upon it without first absorbing the energy.
relative humidity (RH)	The ratio between the actual water vapour pressure of a sample of air at a particular temperature and the maximum vapour pressure the sample could contain at the same temperature, *or* the amount of water vapour present in the air expressed as a percentage of the amount of water vapour that would be required to saturate the air at the same temperature.

saturation point	The point at which a sample of air at a particular temperature can contain no more water vapour. When it is, in fact, at 100% relative humidity.
shading coefficient	An expression of glazing performance, expressed as a fraction of the total solar transmission through clear single glazing.
sol-air temperature	The equivalent of external *environmental temperature*.
solar absorbtivity	The ratio of the amount of solar radiation absorbed to that which is incident on the surface.
solar constant	The amount of solar radiation reaching the earth's atmosphere (approximately 1395 W/m²).
stack effect	The tendency of air to rise from lower to upper openings in the building shell due to *buoyancy*.
standard U values	See *thermal transmittance*.
steady state conditions	Conditions of constant heat input.
sun chart	A diagrammatic method of assessing solar radiation levels falling on a proposed building.
surface condensation	Condensation occurring on the surface of a building element.
surface factor (F)	That proportion of heat gain at the surface of a building element subjected to cyclic input which is re-admitted to the adjacent space.
temperate climate	A climate in which the dry bulb temperature of the air may rise to between 13°C and 32°C and the wet bulb temperature to 24°C.
temperature gradient	A graphical representation of the temperature difference through the thickness of a building element.
thermal capacity	That property of a material or element of structure which expresses its ability to store heat and the rate at which the temperature of the material or construction will alter with changes in temperature of the adjacent air. Thermal capacity increases with density and mass.
thermal conductance (C)	The thermal transmittance between inner and outer surfaces of a construction, depending on the thickness and the thermal properties of the materials making up the construction, but ignoring the resistances of the inner and outer surfaces.
thermal conductivity	The quantity of heat in steady state conditions transmitted in unit time through unit area of unit thickness for unit temperature difference. It is a property of the material which is measured in W/m°C.
thermal insulation	Material of low thermal conductivity used for the purpose of obstructing heat flow.
thermal resistance (R)	The reciprocal of *thermal conductance*. It is a measure of the overall thermal resistance of a material or combination of materials of specific thickness to heat flow. The *total thermal resistance* of an element of structure, including the resistance of the inner and outer surfaces and of any enclosed cavity, is the reciprocal of the total *transmittance*. It is measured in m² °C/W.
thermal resistivity (r)	That property of a material regardless of size or thickness which is the reciprocal of *conductivity* (m°C/W).
thermal transmittance (U)	The quantity of heat that will flow through unit area in unit time per unit difference of temperature between the internal and external environments in steady state conditions. It is measured in W/m² °C and differs from *thermal conductance* only in that it includes inner and outer surface resistances. *Standard U values* are thermal transmittances given certain standard conditions such as exposure and moisture content.
vapour barrier	Part of a construction through which water vapour cannot pass and incorporated to prevent moisture penetration through the construction.
vapour check	A material or part of a construction that offers a high resistance to the passage of water vapour.
vapour diffusivity	That property of a material independent of thickness which is a measure of the rate at which vapour will pass through the material when a difference of pressure exists between the air on opposite sides. It is the reciprocal of *vapour resistivity* and is measured in gm/MN.

vapour resistance	The overall resistance of a material or combination of materials of specific thickness to vapour diffusion in MN/g. (Resistance to vapour diffusion of thin membranes is expressed as *vapour resistance*. That of other materials is their *vapour resistivity* multiplied by their thickness).
'warm' roof	A roof structure with the insulating layer above the roof void.

INDEX